A Paradoxical Life: Where Did We Come From?
By Diondre Mompoint
Diondremompoint.com
Copyright © by 2022 Diondre Mompoint
All rights reserved.
Published in the United States of America
by Diondre Mompoint
No portion of this book may be reproduced in any form without permission from the publisher, except as permitted
by U.S. copyright law.
For permissions contact:
diondremompoint@gmail.com

Cover, design and illustrations by Diondre Mompoint

ISBN: 979-8-9859447-4-7
E-book ISBN: 979-8-9859447-1-6

1st Paperback Edition

METABOLIC PATHWAYS

BY: EVANS LOVE

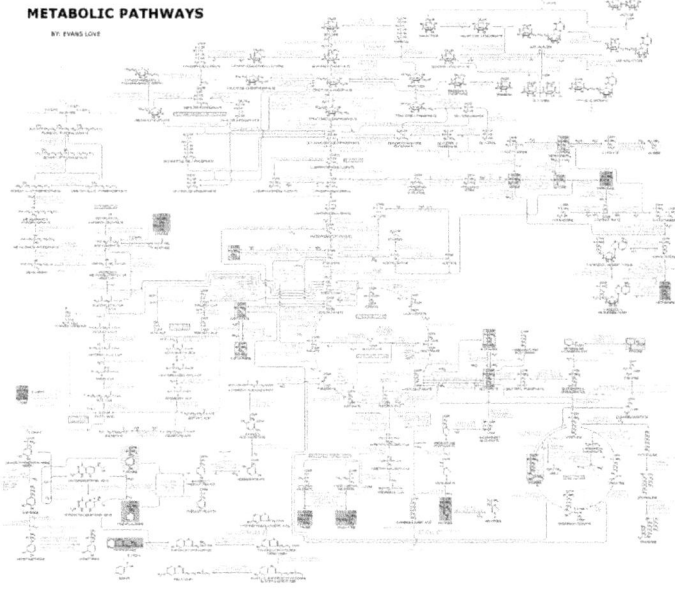

A Paradoxical Life

A Biochemical Approach
Where Did We Come From?

Diondre Mompoint

To My Family and All Future Scientists:
This book is dedicated to my family and all future scientists who embark on the many mysteries science hands us. I want to stress the importance of being fearless and driven when reaching a goal. Before writing this book, I feared that I wouldn't be able to accomplish writing any scientific nor philosophical piece; neither did I think that I would expand my knowledge in the scientific field. This doubt has transformed to confidence and ambition. I want to spread this to all of you! Go write that book that you've always wanted. Go and get that realtor license; open that business you've always talked about.... Again, I share this personal note to all of my family, colleagues, friends, and other innovative minds. Always keep your vibrational frequencies high and your doubts lower.

CONTENTS

Part I: The Paradox of Life

    1. Life is Complex   15

Part II: Let's Open Up a Biochemistry Textbook

    2. Biology is Complex   26

Part III: We Understand How It Works.....

    3. Religion & Spirituality   64

    4. Creationists vs. Evolutionists   87

Part V: Two Boots One Base

    5. The Idea of Collective Origin   103

Part VI: Close The Book

    6. PSSM End   117

About The Author

Diondre Mompoint is an American Biologist from Haitian/Bahamian descent. He grew up in Miami, Florida and was fascinated in engineering and natural sciences. He picked up interest in Chemistry and Biology as a young child and into his teenaged years.

Diondre graduated from Florida Atlantic University in Boca Raton, FL with his Bachelor of Sciences in Biological Sciences. After this, he has spent most of his time studying biochemistry, cellular and molecular biology. He began to question many processes in science as well as the origin of life.

Due to his curiosity and free thinking, he started to form ideas through writing and some journaling. Currently he is working on writing many books for other free-thinking scientists and future scientists. His lifelong goal is to help strengthen the scientific community in academia and continue to shape young and innovative minds through literary work.

Preface

It is imperative that I make a few things clear before starting. For those of you that are a casual reader, I promise to not make this too dense of material. For those that are more advanced when it comes to biochemical terminology, you've reached the right place. Though, I would suggest brushing up on some concepts for that, "Ah-ha" moment when mentioning something like, "branched chain alpha keto dehydrogenase". For those not too fond of scientific terminology, I still recommend reading the first part; I try my best to explain some of these complex processes in a way where others can understand. If your head spins for more than a minute, feel free to skip to the third part where I talk about religion, spirituality and other quirky stuff to feed your brain The one thing that had my head spinning for years are complex molecular machines in biology. They are so structured that I constantly ask myself how is this even possible? If you are not astonished by how biological systems work, then you may not truly understand it. I've spent many years of undergrad learning concepts and having my eyes grow wide as I digest material. After learning how these systems work, there is this dead end. A scientist can learn every single metabolic pathway, but not a single one would be able to explain how it came about. I am the scientist that had this struggle; this is where I am stuck. Any other scientists that claims they do not have this issue may exit stage left, as this book may not be for you. I understand that there are some quantum nerds that can give a fancy explanation, but they still fail to explain how life has come to be. These chemical processes are so specialized that I do not believe the human brain can understand it fully. This leads to the idea that there must be an artist or some creator. The anchors used to back the idea of creation are human curiosity and spirituality that roots from a supernatural being. Followed by this are other possible ideas for the origin of life that have jumped around in my head and are ready to be expressed. To help navigate the discussion and my thoughts, I have organized the book into parts. Before mentioning what each part entails, I warn you of periodical interjections on myself throughout the book; I do this so you can get a feel of what happened as I write. The first part of the book explains the complexities of life. It should leave you with questions that may have you thinking from the beginning to end of the book. By the end you should be able to draft up your own thoughts (literally). The second part uses biochemistry to back up the idea that life is so complex there must be a higher power. In the third part, we can dive into religion and how it may or may not tie into the origins of life. Then we ask how do religion and spirituality differ from one another. After the differences are settled, I explain the subconscious thinking of humans and how that relates to the origin of life. Next, I introduce the arguments between creationists and evolutionists. In this section there will be common arguments made between the two groups. These arguments then get thrown into a rabbit hole about the cosmos, big bang theory and mysteries of the universe. Lastly, the final part closes the discussion with an open discussion about what was discussed before. As these discussions are opened, feel free to respond with an open mind about what you think our origin was. I can see many who are headstrong on science explaining everything. I can also see those who would be surprised that I dare to question Darwinism. It is quite possible to bring Darwinists, Creationists, Evolutionists and all other "-ist" to come to an agreement that not everything is clear-cut. It is also possible that all of the "-ist" can be added into a soup of ideas that represents one theory. Science has reasonable and valid explanation for life yet, these explanations are not always valid for some unexplained phenomenon. I hope after reading this book, minds are opened, and we can all agree that life is indeed a paradox.

Part I: The Paradox of Life

Chapter 1:
Life is Complex

Before you start reading, look around your area and ask yourself what do you see? I will wait……Are there trees, water or animals? Are you reading this at a bus stop with strangers or maybe laying on a grassland with a view of the sky? It would be a terrible lie if you told me you were not fascinated by what you see. From the smallest organisms seen on a microscope, to the mega Hubble Space Telescope that gives us the beautiful image of the Andromeda Galaxy, life is filled with an array of spectacles that leave us curious. With all this information we take in this visual world, is our life really defined by our perception of what we see? The "what is life?" conversation is very broad and can be interpreted differently from one person to the next. The one thing that is the same for each person is the fact that not everything in life truly makes sense. We struggle to give a definite answer for life's meaning. We can sit 1,0000 people in a room and ask them what the purpose of life is. To no surprise, we will get varying answers and a lot of responses like, "live your life to the fullest". I'm tired of these answers since they lack a detailed explanation of what the purpose of life is. There must be more to it. Many people say the most straight-forward answer is usually the answer.

However, this situation is different. Everything really isn't always black and white.

Let me land an example here. As I am writing this, about exactly 5 minutes ago, my cat jumped onto the kitchen counter and knocked over a wine glass that I had drying. I was in the bathroom at the time (do not judge me because I write any and everywhere) and heard a loud noise. I knew that something had broken. Without going into further detail of how I escaped the bathroom (of course I wiped first), I got up and saw the wine glass shattered and other cups that fell into the sink. Remember, I didn't see when he jumped onto the counter to even knock it over. Since I couldn't see, am I going to pace back and forth and try to figure out how this happened? I can say it was:

1) A ghost (not as absurd).
2) A wind slipped through the crack of my door and tipped the wine glass over.
3) Someone in another country got shot in the foot. Out of anger, they smacked a wine glass and every other wine glass in the world was hit as well.

These assumptions would be absurd and illogical. The obvious conclusion here is that the wine glass breaking was from the actions of my cat. We were the only ones in my apartment. I did not have to do any further investigation or thinking for what may have caused this. If only answering the question of what knocked my wine glass over was as easy as answering what is/created life. The explanations and assumptions are far less obvious. Some people reading this probably think that I am digging too deep into this. I see no issue here, since for the longest of time we have been stuck in this paradoxical life. I say we can think outside of the constraints of what we are taught or conditioned to know.

Dictionary Lesson 1

We live in this small bubble of a world that we have yet to fully understand. There are many mysteries and questions to life that are still unsolved. Did you know that only about 5 percent of the ocean has been discovered? Or that we still do not have a strong idea for why animals and humans sleep? Before I get bashed, please neuroscientists do not shake your heads like you know this is false. If you want to discuss, please email me. If we move on, how about explaining mysterious astronomical objects that fly across the sky? These mysteries have some of the top astrophysicists running out of chalk and board space!

We must face the truth; understanding life and everything in between is complicated. So complicated, that some of our explanations are contradictory of our beliefs; it is paradoxical. In the preface I promised to hold your hands throughout the book when too many complex terms are thrown around. So, before I jump too soon, let's play a little dictionary game. There will be a few more dictionary games like this throughout the book to help you all follow along. If you feel you are too smart for these baby definitions go grab yourself a cookie and pat yourself on the back; I guess you're too smart. A paradox is a statement that contradicts itself; it is both untrue and true simultaneously. A type of paradox or synonymous with paradox is antinomy. For this book, we use antinomy since we discuss two theories that contradict one another (evolution & creationism). I don't want to get too caught up with all the other paradoxes; for our journey we are keeping it simple. We can leave it as believing one thing is true, but also may be untrue to the mind. The most common paradoxes have been seen in some in some of Shakespeare's writing and other famous creatives (hopefully me too). Here are some examples of statements that can help you get familiar with what a paradox is:

1) This is the beginning of the end.
2) I'm a compulsive liar.
3) I must be cruel to be kind.
4) Waking is dreaming.
5) If you don't risk anything, you risk nothing.

6) Less is more.
7) *You're damned if you do and damned if you don't.*

Above, are contradictory statements that are common. For example, let's look at the one that says, "You're damned if you do and damned if you don't". This statement is my favorite since it is something that my mother would always say. Who knew that something she would always say would end up in my book? Hi mommy! I never thought the saying was a paradox, let alone know what a paradox was. This statement is contradictory since the person in the situation gets blamed regardless of the outcome. If I wash the dishes, then I will be yelled at for not taking out the trash by my mom. If I don't wash the dishes, I will still get yelled at for not doing it at all. In this scenario it is self-contradictory and the same outcome. I feel the same contradiction when being in the middle of science and religion/spiritual beliefs. I know that science has more evidence against many of these "mystical beliefs". I find myself in this life paradox where I question anomalies and expect science to explain it. Then I relapse, to say, "There is no way science can explain this one". Or in other instances I say, "there is no way this could have happened and cannot be explained by the faith". I'm sure many of you feel the same way but may not want to burst out and say it.

The Mompoint Paradox

With life's paradox and the contradictory mind that I've had for the longest, I present the Mompoint Paradox. Before I go more into detail, it took me a while to decide what I should call this paradox. Looking at some of the other boring names I decided to use one that makes the most sense to me, my name. I know it's a bit egotistical to name it after myself, but isn't that what all the other cool guys do? I want to make this paradox personal and something that can be used to represent others that feel the same way. The Mompoint Paradox says that science explains everything, but everything cannot be explained by science.

A few things that I want to make clear about this paradox is wording. When I say science explains everything, I did not specify whether that explanation must be 100 percent correct. To explain something means to make an idea more clear and present facts; science does just that. Where science doesn't do this is when ideas are made clear and only some facts, not all, are presented. Whatever is deemed to be inadequate factual evidence can be questioned through principles outside of science.

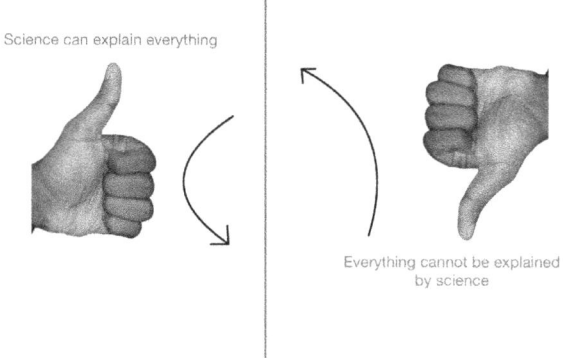

Another rebuttal to this paradox is that we cannot say science can't explain everything just off the mere fact that we do not understand. Even if we did understand everything the universe has to offer, what would that even be like? The answer would still be that nobody knows. In the scientific minds of many, it is hard to even fix your mouth to say, "I don't know". Our minds get clouded by wanting to believe we know everything. We have a strong sense of what we think is correct. I can literally hold up a picture of the Earth and provide all the facts and evidence for a spherical Earth. There would still be individuals that believe in a flat Earth (scientifically impossible). Who am I to go back and forth with someone that believes opposite of what I think? In no way shape or form am I agreeing with a flat earther, rather, I am allowing for them to freely believe what they believe. The Earth must be spherical in order to orbit around a larger mass (the sun) and gravity would be meaningless. I can go into further detail for why the flat earther would be wrong, but then they can slap my own paradox right in my face. They can say, well Mr. Mompoint, you said, "science explains everything, but everything cannot be explained by science. Therefore, your explanation for Earth being spherical is invalid." These clouded, "we know everything" moments have created these arguments that we have today: The arguments of climate change, what is the right vaccine to take and the right answer to the origin of life.

Am I A Believer?

There is this constant battle between science and religion. I continue in this constant limbo between the two and finally have the courage to express the ideas that make most sense to me. I call this a middle ground or a way to make myself and others feel at peace. We should no longer have to have a foot in both camps. Instead, we can merge our camps together and become one. I like to start off saying that I am a firm believer in God, spiritualty and engagement in prayer. I also enjoy and understand biological concepts that try to explain the origin of life. So wait, I am someone that believes in a God, but also accepts scientific principles? Well, allow me to clear this up. I am a supporter of some parts of religion and the idea that there is a higher power. When we speak in scientific terms, we are told we must accept the logic in the theory of evolution. So, where does a scientist that believes in God and respects science stand? It is possible to believe in God and understand that God created "science". WOW! Yes, I know this is a bold statement; I actually intend on using this statement to open up the discussion that life is so complex that creation theory is not to be ruled out when comparing it to Darwin's theory. Throughout this book you'll see me use creation, design and intelligent design interchangeably. These are the different ways of saying we and the universe were created by something. Before you close the book and think this is all mumbo-jumbo, please stay for later chapters that will help explain this. We often ask why do humans act the way they do? How did we come to be? Science explains this by using Charles Darwin's evolution theory. In simple terms, this is the idea that humans and other living organisms have evolved from a previous species. It also expresses many genetic mutations that occur to help make that species more fit for its environment.

There are many pieces of the pie that Darwin's Theory of Evolution cover, however I promised to keep this simple...for now. After digging into his theory, I understand assumptions would be made for an organism developing characteristics that make them unique. It all makes sense at first, but there still is an unknown.

In the grand scheme, Darwin fails to give the answer of why we are here. I'm confident his observation and studies were not to answer the "why", but at least explain "how" we came about. I can say that not a single person can answer why we are here. However, how we got here is the debate. The scientific explanation for why we are here is to survive and reproduce. Is that really all? The complexities of life should serve a greater purpose. Shouldn't it? At least a purpose for the person that created us in his/her/its own eyes. Are we just some chess pieces on a quintillion-by-quintillion sized chess mat being played with our lives? Some may say that we are in some simulation (I don't want to go too far here, I may attract the wrong type of reader, "cough" quantum geeks)? Hate to say it, but others may say our lives are just meaningless; we are just here. If we move away from the philosophical ideas of life, science is much more straightforward (sometimes).

Part II: Let's Open A Biochemistry Textbook

Chapter 2:
Biology is Complex

Out of all the branches of science, biology gives us the foundation for all of life. Many unexplained phenomena were debunked through scientific studies in biology and its other branches. Biology explains the natural processes of life, how organisms grow and reproduce sexually. Of course, there is more to biology than just sex you horn- dogs; it also has subjects involving human behavior, reaction to environment, ecosystems, organisms' anatomy and more. Even if you haven't picked up that dusty Biology book from high school in a while, I'm pretty sure many of us understand these broad topics in biology. I want to focus on one topic that many should be familiar with; that is the cell structure. I hope all you kiddies remember those fill in the blank test you've received in middle school. We had to fill in some poorly shaded diagram of an animal cell or plant cell and usually got points deducted because the structure was not identifiable. If you're not too familiar with the animal cell, we may refresh our brains here. The cell structure is one of the first anchors that I will use for my overall idea. Through decades of study, scientists have learned the intricacies of cellular structure and function. Before this, the cell was thought to be just some glob of plasma.

Today, we know that this glob theory is simply untrue. In a general biology course we have learned that all living organisms are made up of cells. These cells have organelles that serve SPECIALIZED jobs that help the cell function (I hope all this is coming back to you guys). When these trillions of cells function together they help form tissue, organs, an organ system and the organism. We can think of the cell as the most efficient production factory; something like Amazon (yes, the cell doesn't get bathroom breaks either). In the cell we have the nucleus where instruction is given (command center). Before I continue, I understand the concepts are basic for some. Please bear with me. I will excite you biochem geeks in the next part (sorry pure chemists). Let's continue; the ribosome (assembler) assembles protein that will be used around the cell for other functions. The rough ER (transporter) transports the ribosomes that make protein. The golgi apparatus (packager) packages and ships proteins to their proper destination. Lastly, we have the mitochondria (energy source)! Most famous for its very dry definition: "the powerhouse of the cell". It is so much more than that; in simple terms it provides and produces energy for the cell. The figure below shows a beautiful and distinguishable animal cell with its parts labeled. I would say it is so beautifully drawn out that your middle school teacher probably would be jealous.

Your middle school teacher and others all agree that the cell structure is special. Funny enough, we are still far from the complexities I would like to share (I hope you laughed). I use the organelles as one of those anchors that I've mentioned earlier. The cell is highly efficient and an organized processing facility. Their inner workings resemble a unique model that must have been created by someone or something. All of the cellular organelles are moving parts. One may not work if the other is not present. This is introducing the idea of irreducible complexity; this is a term coined by Dr. Michael J. Behe. His main anchor was explaining the bacterial flagellum and how it would cease function if a protein was removed from its motor. Well, here I use the same idea, however at the cellular level. Evolution Theory says that many of these structures have evolved over time. This means that some parts of the cell may have never been there until it evolved from something else. Now kiddies, we know that the nucleus works in conjunction with the ribosomes and golgi and so on and so forth. I believe it is best to say that the formulation of these structures may have been pre-designed pieces that fit together. I stress the words, "may have been". Removing any piece of the structure would make the entire cell factory cease function. Imagine having the nucleus (command center) removed from the whole cell? Instructions for the cell and vital proteins would not be produced. We may consider microevolutions in the chemical make up for these structures, however even that would not explain the timeline for the formation of the organelles.Let's say if there were some microevolutions overtime, then where would the previous parts have derived from? I really want everyone to hold this point as this is just a little foreshadowing for later chapters. Another explanation could be that the nucleus has always been there, but one part of the nucleus did not arrive until later. We can think of this like a person who owns a home but no air conditioning (use your imagination here, I could not imagine someone purchasing a house without AC). Let's call this person, "Darwin". So everyday Darwin uses a stand-up fan and sweats his days away during a terrible heat wave. He works at a factory that manufactures religious books and is miserable each day. It is not until one day he found out about an AC unit that has been stored away in the basement. He had no idea it was there, but it has always been there since he has purchased the home. At this point, Darwin decides, "Hey I have no AC and I can really use this to make my days easier; I will be less irritable.". Darwin is excited to use this unit to enjoy some cool air, but there are some missing bolts, screws and a blade. It isn't until this moment; Darwin orders all the pieces needed, and by the next week has a repair guy come in to fix it. The following week he can use the AC and reads his favorite book, "A Paradoxical Life: Where did We Come From?". This example gives the idea that the nucleus has always been in the cell but may have not had the pieces needed for it to function or was not used until it was needed. Could the cell had said, "wow there's a nucleus here that I didn't know about. I could really use this thing! Let me make it work!" Again, these are all just rhetorical questions to make you guys think. If we dive even deeper, we can see the complexities of DNA and how it serves the idea of a creator. We can revisit the analogy that was stated before; the nucleus is like the command center (CC) for the factory. In this CC we have these computers that code, more specifically, for pre-determined functions in the cell. These series of code are called DNA. The DNA is responsible for carrying genetic code that will produce proteins, initiate cell replications, growth, and determine how an organism appears (thank your parents for this). The DNA is then comprised of nitrogenous bases: Thymine, Alanine, Guanine, Cytosine and at times Uracil. When these nitrogenous bases are paired in a certain way, this gives instruction to create these amino acids. Amino acids are the essential precursors to proteins. One of those other boring definitions that you learn, "the building blocks for proteins". The amino acids then become long chains called polypeptides and eventually form a functional protein. The DNA doesn't just stop there; it is a way cooler kid than this. DNA can also be triggered to produce proteins in response to something else (I will touch more on this later I pinky promise).

TACGATGTCGATCGATGGGCTATGCG-TAGGCTAGCTGGGATA

We can compare DNA and its ability to drive function to codes on a computer software. I am pretty sure many of you are reading via a digital book or some sort of app. These apps were created by computer programmers or software engineers (wink wink). Programmers use certain codes for function when an action is taken and usually help maintain the design/overall function of the software. I know that computer engineers and software programmers may be different, but I shall use them interchangeably for making the point. Here is a chart for some guidance:

Creator--→ Programmer
DNA----→Computer Software
Function---→ Code

Before I continue, allow me to add in this disclaimer: I am aware of the debate or argument that DNA is not like a computer since DNA code is random, and computers are algorithmic etc. I refute this claim and say that DNA sequences and mutations are random, but the overall construction of DNA is not random. I would like to cut out the logistics and stick to the basic idea that DNA and computer code are not 100% the same, but similar. To also help, please look up the definition of "similar". I am a bit excessive and over the top when it comes to supporting the overall idea. So, in these next few paragraphs, I will leave the word, "similar" and "similarity" in CAPS for extra emphasis. Now that we have that squared away, I may continue to speak about the root^2 of this all (see what I did there).

Allow me to drop an example here. We are all familiar with many of these online shopping websites and how they operate. You first open up a web browser and type in the unique website URL. You shop around and find some nice jeans, a graphic t-shirt and shoes. When you click the "add to your cart" button, a set of code is triggered to show a "1 or 2" indicating the amount of items that are in your cart. When you click the cart button, you are redirected to the checkout page and eventually instructed to type in your credit card information. Code can even go as far as initiating those pop-ups that say, "sign up for our newsletter" and have site redirection links or interactive widgets. All websites are an interconnection of triggers, switches and code that activate one function to another by clicks. Each code is uniquely created to serve a function; the same is true for DNA. There will be more anchors about this whole trigger/switch comparison in the next chapter.

  Let us also revisit how software engineers maintain these websites. Why do I bring this up? With all the websites and apps there are constant updates because of bugs (bear with me here). A bug is an error in coding. Often times this happens when there was a mistake made in the program's design or source code. A software engineer is responsible for maintaining these codes and fixing them when needed. Overall we understand that some star-studded computer programmer designed all of these interconnections. The way DNA code was created is very SIMILAR to how computer software engineers create computer code. DNA has codes that are composed of nitrogenous bases. When they are lined up in a specific way they code for proteins that will be used for function within the cell and the body. SIMILAR to software programs, some DNA codes are triggered by responses in the cell. Just like any computer code, DNA has its own bugs too. These bugs are called mutations. Mutations occur in DNA when there is a change in sequence through copying mistakes and infections or diseases. SIMILAR to software engineers, it is up to DNA repair proteins to maintain and repair the genetic material. Again, staying away from the detailed specifics, we understand the SIMILARITIES between DNA code and software program code.

    I want to loop back around and really tie this all together. So far I've talked about the overall complexity of life explained by general biology. Next, I took a dive into the cell and all of its organelles and functions. Then I mentioned DNA and its own complexity. Now we can dive into some chemistry (I'm sure all the chemistry fan boys skipped here). Again, it will be very basic concept to support the overall thesis. I'm sorry we won't be talking about synthesizing maitotoxin or some crazy mechanism for the formation of i-propyl cyanide; we want to stick to the basics for now.

If we look at what makes up cells and DNA, we will see that atoms link together to make compounds and molecules. These atoms are uniquely arranged to form structure proteins, cell walls, hormones, signaling proteins, enzymes, and many other important complexes. The main molecules we have that are the foundations to life are carbohydrates, lipids, fatty acids, nucleic acids and proteins. The molecules mentioned are guided by other molecules and usually do not know where to go. They are highly specialized. Some require optimal pH levels, change in temperature, high or low concentrations of a certain molecule and energy. The specification is out of this world! Molecules that are found in enzymes for example are usually directed by other amino acids (we will see an example in gene regulation later). There is high specificity for some molecules to be together; this is far from being random.

In any general chemistry course we learn about how elements react with one another. My question is how do these elements readily react with one another? No, I do not want a physical chemistry explanation either. I bring this up to highlight that many of atomic behavior seems to be programmed as well. Whether the reaction of two atoms is spontaneous or not, they just know what to do. Is there really ever a "bad mix" or "bad reaction"? Sure nobody would want to stick his or her face in front of a beaker when mixing ammonia and bleach, but it is still considered a reaction whether it is good or bad. These reactions are structurally created through stepwise mechanisms.

$$\underset{\text{Sodium hypochlorite}}{NaOCl} + \underset{\text{Hydrogen chloride}}{2HCl} \rightarrow \underset{\text{Chlorine}}{Cl_2} + \underset{\text{Sodium chloride}}{NaCl} + \underset{\text{Water}}{H_2O}$$

$$\underset{\text{Ammonia}}{2NH_3} + \underset{\text{Chlorine}}{Cl_2} \rightarrow \underset{\text{Chloramine}}{2NH_2Cl}$$

Compounds spontaneously come together as if they were meant to be future couples. It is genuinely a beautiful love story. Sodium hypochlorite (bleach), ammonia, chlorine and water mix to form chloramine; a toxic gas that causes nausea and shortness of breath. I guess it isn't always a beautiful love story. Whether the reaction is ugly or beautiful, these atoms form bonds regardless. A more practical example that relates to something in our everyday life is the formation of the structural protein alpha-keratin. This protein is formed by amino acids that link together by peptide bonds. The protein is responsible for the growth of hair in vertebrates and even nails. There are repeats in amino acid residues that bond just like that future couple I talked about before. The carboxyl group reacts with the amino group of another amino acid. The formation of the peptide bond is an unfavorable reaction that requires plenty of energy. These chemicals are very far from dumb and don't like to waste energy. They are very sassy and would only do work if it is favorable. To overcome this, the enzyme peptidyl transferase lowers the activation energy for the formation of these bonds. Now that less amounts of energy are needed, the molecules can form that bond. Lastly, water and heat are released; the overall reaction is called a dehydration reaction.

Peptide Bond

It is all like a speed date! We have these amino acids that want to date and be together. Then we have a host (the enzyme peptidyl transferase) setting up these dates and lowering the nervousness of these hot and ready amino acids. The host facilitates these speed dater's ability to bond. Once the nervousness is lowered and the amino acids get comfortable, they can link together. All the heat and water (sweat) are then released. Now this is a true love story! The formation of bonds and large macromolecules become perfect matches for functions throughout the body. They are perfectly designed, and I would even say...programmed (**giggles**).

If we move our way back, we've discussed complexities at the atomic level, molecular level and cellular level. We haven't even discussed the subatomic nor quantum mechanics' level; the levels I have mentioned are just enough to highlight the complex systems in life. The purpose of this chapter is to introduce some of the basic biological concepts that would make any scientist scratch their head. Again, not that we would be scratching our head about how things work, rather how super sophisticated these biological systems are. As I am writing this now, I am scratching my head... Never mind, could just be the dandruff. How did that come to be?

Let's Get More Complex!

Fair warning! This part will be dense! Go get your biochemistry book or borrow one from a friend of yours. For those that may not understand biochemistry and are a casual reader, you may tag along if you like. Another option is to skip to the spirituality and religion part then read onwards. If you stick through this chapter, I truly commend you! I want to open up this chapter with introducing an analogy talked about in the previous chapter to help readers get an understanding of the anchors I will discuss. I've used the words, "triggers" and "switches" to describe basic concepts in general biology and chemistry. There will be a clearer reason for why I use these words to describe chemical processes and molecular machines. Remember that computer code that we talked about? Well, yeah that! Computer code isn't just a good example for peptide bond formation and cellular operations. Many switches and toggles are seen in the study of biochemistry too. We are literally talking about biological systems and explaining it using chemistry. I've never met a greater match than these two subjects. Though this subject probably isn't everyone's true love story, biochemistry is a tad bit different. It is no longer this, "powerhouse of the cell thing" or "central dogma". We are talking about the nitty gritty; the stuff (I would rather another word than stuff) half of your lecture hall fails in undergrad. As a lover boy for the subject, I can say it is hard! It is hard because of the complex structures and the fine detail for how these mechanisms work. I always want to reiterate the main idea here. Biochemistry helps us understand how life works in a more detailed fashion. I'm guessing many of you probably beat me to what I was going to say next! Yes! There must be a creator! I am going to outline this chapter with a series of biological systems that I find most complex. It will tie into this whole paradox created by life and help those see where life's processes may contradict what we think we know. Before I ramble into question, I would like to thank everyone that is still reading at this point! I digress. Here we go! All the topics that I touch base on will leave questions for some scientists. Why is it that processes are so detailed in the metabolism of molecules? If we are supposed to be as efficient as possible, wouldn't it make sense for some of our biological processes to be simpler? A part of Darwin's theory of evolution states that changes occur for the species to be better fit for its environment. The multiple insertions, deletions and mutations within a species genome cause more complex structures to be formed (at the molecular level). So wouldn't complexity make life harder than being beneficial? I've looked at over hundreds of paramecia and diatoms under my microscope. These single celled organisms seem just fine to me. They are very simple compared to any other species. How come we could not have stayed that simple?

We just had to have this sophisticated multicellular universe of our own. How about species that use light energy to process food in photosynthesis? An inorganic photon of light to make organic matter. Programmed Cell Death and the caspase cascade? These are questions that stumble scientists including myself. The anchor I want to start off with is gene regulation and its specialized system of switches. I am very familiar with the regulation of genes from when I was born; I am hoping all of you would say the same.

The reason I say this is because as we grow from being an infant, special regulatory proteins transcribe many genes. Proteins guide cells that are needed to form your little hands, your head and reproductive parts in the womb. That is why we can use ultrasound and say, "yay, it's a girl" when determining the sex of our children. These proteins can rapidly influence gene expression or halt it on a dime. Cells become ordered and structuralized into skin cells, liver cells, heart cells and so on. These cells are called pluripotent cells. This means that the cells have the ability to differentiate into any of the three germ layers: endotherm, ectoderm and mesoderm. Many of the systems that we know arise from the germ layers mentioned. This includes the nervous system, skeletal and muscle system. So, how do we control these behemoths that have this power to transform into a variety of cells and huge functional systems? Well would you look at that? I was beaten again! Yes, gene regulation plays the part on controlling how these cells differentiate in the growth of organisms.

Gene Regulation

All eukaryotes experience gene regulation in some form. I agree we all understand how our cells were being directed into what their fate will be (i.e. liver cell, heart cells etc.). Allow me to bring in a more practical example. We shall set the scene at a party where you've just found out you've been promoted. There is fanfare, shouting and bottles being popped. By the end of the night you may have had a bit too much to drink. With this, you now have trouble standing up, talking and functioning. Some of the drinks you've consumed had alcohol! The alcohol compromises the central nervous system causing impairment and poor judgment. I am a Biochemist for God's sake (Darwin's sake for some hehe), so let's cut to the more in-depth stuff. As alcohol enters your system, it is processed through the liver (which arose from those pluripotent cells guided by gene regulation **wink**). When ethanol enters the liver, gene transcription is TRIGGERED. The genes involved in the formation of enzymes to metabolize (break down) alcohol are ADH1A, ADH1C, ADH4, ADH5, ADH6 and ADH7. All of the following genes are fairly similar but differ in a few properties. Once transcribed, we form the Mack-daddy himself! We will have alcohol dehydrogenase (ADH1C CAS 9031-72-5). This is a zinc enzyme that will convert the alcohol into a ketone or aldehyde. In this case however, the ethanol will have a high affinity to the ADH and bind to the active site. ADH oxidizes ethanol to acetaldehyde and reducing NAD+ to NADH. The majority of enzymes involved in this interaction occur in the liver. Other reactions may occur in the brain and other parts of the body at much lower rates.

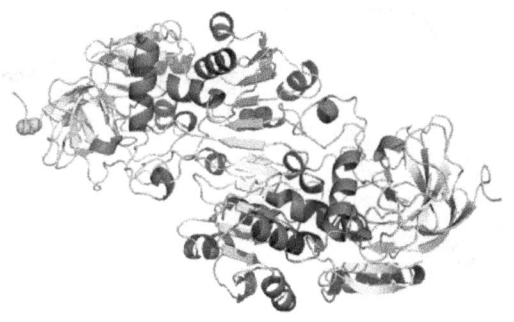

Alcohol Dehydrogenase

When there are very high levels of consumption for ethanol in the body, the Microsomal-Ethanol Oxidizing System (MEOS) kicks in. This serves as an alternative to the metabolism of ethanol, which occurs in the cytoplasm of the liver cell. The MEOS occurs in the endoplasmic reticulum respectively. This pathway requires a different enzyme called CYPE21 as a part of the cytochrome P450 enzyme family. In both the regular consumption and high-level consumption, the conversion of the ethanol is important. When you wake up from that celebration and have that hangover it is because of that toxic acetaldehyde hanging around. The body readily converts it into acetate using acetaldehyde dehydrogenase and the conversion of NAD+ to NADH.

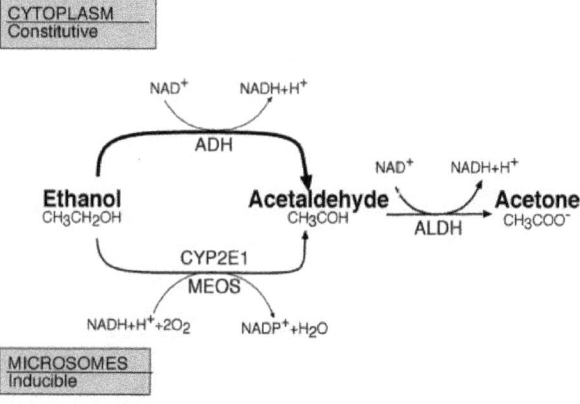

In the final formation of the acetate, it is then used in the body to regulate appetite, hormones and other metabolic pathways. That "celebration acetate" will now become part of some other metabolic pathway used in the body. You see what I mean here? The conversion of ethanol to some other molecule used in the body is another great example of a complex system that is irreducibly complex.

Okay guys, quick intermission here! I hope some of this regulatory and metabolic pathway set the stage for a complex life. Let us think about this again without all the big words. We literally have a system in place to break down everything into something else that will be used for something else. If we were to remove alcohol dehydrogenase from the metabolism of ethanol, then how was it broken down before? It is an irreducibly complex system (inserts Michael Behe's smile here). I figured there would be some debate over this idea. So allow me to beat you guys to the possible explanations:

1) Alcohol dehydrogenase has always been involved in the metabolism of ethanol, however it may have been a downgraded version of the ADH protein we know now.
2) Alcohol consumption was not as prevalent in our ancestor's past. Therefore, the MEOS and breakdown of ethanol was not around.
3) The system had to have evolved from another similar system that breaks down something else. There must have been multiple precursors to the ADH. After many generations, the genes used to transcribe ADH have changed overtime.
4) Quantum (this is a joke guys).

I hope I beat you debaters to what some possible explanations are. I would give rebuttal to all 4 individually; however the response I have should answer all of them.

The main idea of the 4 explanations is that ADH may have evolved from a previous metabolic pathway or was once a simple enzyme upgraded to a zinc enzyme. This poses the question of where did that simple enzyme come from and the enzyme before that and even before that? At some point, where do we say something was created? It seems like I end each of these topics with a question; this is the classic answering a question with an answer then with a question. I believe these questions say a lot about what we don't know. However, please hang tight as I have a possible explanation in the later half of this book. If this wasn't enough complex "you know what", then I would like to mention we are just getting started. So, biochemists, chemists, molecular biologists (no quantum geeks) and whoever else is reading let's continue.

    I want to talk about genes transcribed in prokaryotes. More specifically, the Escherichia coli found in the intestines of living species. Though bacteria and unicellular species are fairly simple, they too hold very complex systems. E. coli and humans share a symbiotic relationship; so next time you eat please rub your stomach and say thank you! These bacteria are essential to the micro biome and break down the foods we eat. These little guys are fairly harmless and are important to a healthy digestive system. E. Coli has a unique sequence of genes that are responsible for the breakdown of lactose into simpler sugars. This whole operation is called the lac operon. The lac operon was one of the first examples of gene regulation that I learned in my undergraduate program. The main components of the operon are the CAP site, operator, promoter, the LacZ, LacY, and LacA genes. The lac operon is considered an inducible system, which means that it is usually turned off and turns on only when there is an inducer that appears. This is the trigger system I was talking about. The operon is only going to transcribe genes when needed; Oh yes we have another sassy system here! When lactose is absent, a repressor protein binds to the operator of the gene sequence. A repressor is a protein that silences genes from being expressed. The operator regulates the production of genes for a certain sequence. The operator overlaps the promoter where DNA polymerase binds. Since the repressor is bound to the operator, the RNA polymerase cannot bind to the promoter and transcribe the upstream genes LacZ, LacY and LacA (all genes responsible for breaking down lactose). The Lac Z transcribes the protein β-Galactosidase, which cleaves lactose into glucose and galactose. The LacY helps transport the lactose into the cell and the LacA is a trans-acetylase. The LacA is not too much understood at the moment; in simple terms nobody knows what it does. Side note: Imagine what other things we do not know or fail to say we don't know. When lactose is present, an inducer called allolactose binds to the lac repressor. The allolactose shoves the repressor away and forces it to leave the operator. This TRIGGERS the metabolism of the lactose and the opportunity for RNA polymerase to transcribe the genes.

This process happens at a very slow rate, so to boost the process another complex structure joins the action. C-AMP-CAP complex speeds up the rate of the reaction by lowering the activation energy (E_a) and promotes the RNA polymerase activity. The promotion of transcription is in the scenario when there is lactose present and no glucose. The c-AMP or Cyclic Adenosine Monophosphate is rich in the cell and binds to the CAP (catabolite activator protein). With the binding of the cAMP and CAP, they form the c-AMP-CAP complex. This complex then binds to the CAP site of the operon allowing for crazy amounts of transcription. Whew... I warned you guys this would be dense!

When there is lactose and glucose present, the same scenarios apply. The only difference here would be that since there is glucose present, there would be lower amounts of cyclic AMP. With lower amounts of cyclic AMP comes less transcription. The CAMP will not be able to bind to the CAP; there will be no formation of the

Where Did We Come From? 49

c-AMP-CAP that will boost transcription. For the more visually inclined look at the diagram below. I told you guys I promised to try to hold your hand here; hopefully your palms aren't too sweaty yet.

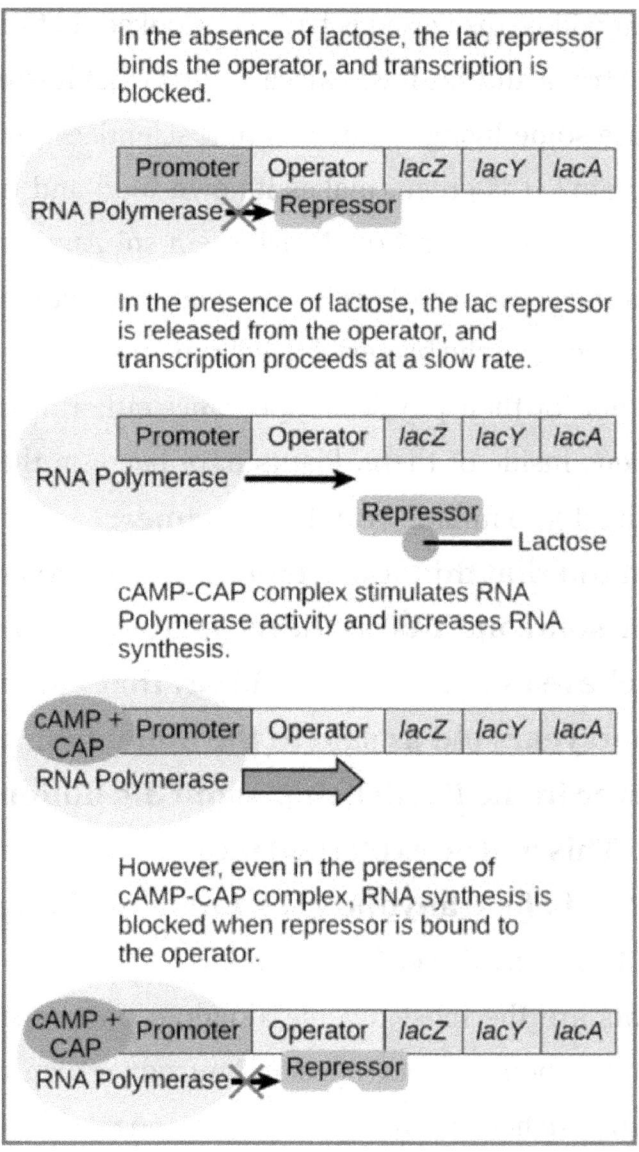

Our Green Friends

Yet another complex structure shown to you spoiled readers; we are halfway through so hang tight. The next process that will be discussed is truly overlooked. Let us play a bit of trivia and see if you all can guess what it will be. What are some living organisms that scientists still cannot explain? It is green, makes its own food and is used as décor. Plants! Oh how I love them so! Anyone that knows me says that I have a very high fascination for plants (second to the bacterial flagellum and the cosmos). It is not for their physical appearance either; it is what happens inside of them. Plants have been on the earth for about 500 million years! **Disclaimer: [I understand that this statement may be contradictory of someone that believe in God. Those who do believe in this Deity would say that Earth is just 6000 years old as said in the bible. I am a sole believer in the Earth being about 4.5 billion years old. This will be explained more in the next chapter. So, before anyone goes nuts, just hang tight and hold hands with a plant].**

Plants are the most diverse kingdom on planet earth and total about 391,000 species. I introduce plants to the equation to help many of you understand the overall point here. Animal lives are not the only ones that are complex. Plants have a lot of weird stuff going on too. When I say life is complex this includes human life, plant life, animal life and life that may be beyond our scope. Though the main highlight of this book is the complexities of human life, I would like to include other organism's life to drive the point. Can anyone guess what the process will be? I will just leave one word here: Photosynthesis...

Where do I even start? I usually get very aroused when hearing this word. Why say that? Oh thank you for asking! Well, photosynthesis holds a dear place in my heart. It is a beautiful, yet a very head scratching process that is not fully understood. This is the only head scratching I can distinguish from my dandruff. Sure, some of our favorite textbooks show some colorful cartoon schematic, but there is more to it. Photosynthesis is a process plants use to make glucose and release oxygen to the atmosphere. It involves water, sunlight and CO_2 produced by animals.

$$6CO_2 + 6H_2O \xrightarrow{Light} C_6H_{12}O_6 + 6O_2$$

Before I get into the funky reaction stuff, let's allow some time for a quick summary here. In photosynthesis, photons of light strike the plants leaves containing chloroplasts. In the chloroplasts are chlorophyll molecules that reflect green light, hence giving plants that beautiful green color. The chloroplast also holds thylakoid stacks that take in the light energy. These thylakoid stacks are surrounded by colorless fluid called the stroma. The bottom of the plant has stomata; small holes that partake in gas exchange with CO_2. Lastly, let's not forget that H_2O. Water is taken up by the roots of the plant and used in the splitting of the molecule. When we band all this together we have photosynthesis. Let's start the fun stuff now! COMPLEX! COMPLEX! COMPLEX!

    Photosynthesis can be broken into two separate reactions. The first reaction is called the light-dependent reactions (light reactions). In this reaction, oxygen, ATP and NADPH are created. In the membrane of the thylakoid are photosystems II and I. The chlorophyll in PSII (P680) absorbs the light energy making electrons very excited. Let us remember that, though it is called PSII, it is the first step of photosynthesis. Scientists just named it this way since it was discovered first (**rolls eyes**). The electrons excited by the light then flow through the thylakoid membrane making the membrane negatively charged. Along the membrane are special proteins that aid in the flow of the electrons; this is called the electron transport chain (ETC). The electrons are transported to electron acceptors like plastoquinone (PQH2-→PQ), cytochrome and plastocyanin. As the electrons move through the ETC, they TRIGGER the activation of protein pumps that move H+ ions into the thylakoid. In addition to the hydrogen ions, enzymes split water molecules and allow those hydrogens to enter PSII. The water molecule acts as a replenishment system to the H+ ions lost in PSII. Hmmmm... maybe some of you should stop reading for a bit and go water your favorite plant. Those hydrogen ions are not going to replace themselves. Okay, so hydrogen was broken off from H_2O for the replenishment system so how about the oxygen? Well, the oxygen is released as waste; yes that good ole oxygen we breathe.

In PSI, photons of light excite electrons just like in PSII. The electrons flow from the plastocyanin and then to another electron acceptor called ferredoxin. This small molecule is reduced and continues to allow the electrons to pass. In the stroma electrons now aid in the bounding of H+ and NADP+ to form NADPH. As this is done, let us not forget about the high concentration of hydrogen that is building up in the thylakoid. Those H+ ions will diffuse into the stroma through ATP synthase. In this process, H+ helps bind ADP and a Phosphate to form ATP (Adenosine Triphosphate).

Phew! Okay that was a lot, but let's hang on for why ATP and NADPH are important later on. I also hope you guys are still with me here. I promised myself to really go into detail. Not trying to be a walking textbook (well in this case a writing textbook hehe), but it is important to really open the eyes of some. This stuff is beyond complex! Who would've known all of this goes into self-feeding a plant!

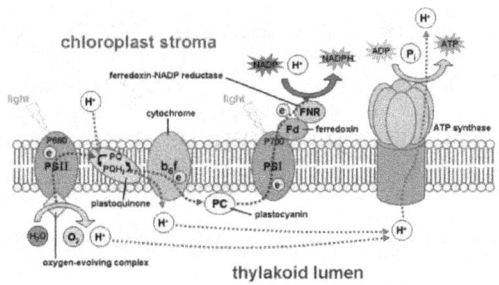

The next part of photosynthesis is the light-independent reactions (dark reactions/Calvin Cycle). In the stroma is a large molecule called Ribulose bisphosphate (RuBP 5C); it bounds with a carbon dioxide molecule making a 6C molecule. The NADPH and ATP from the light dependent reactions I talked about earlier will then be used to form two three-carbon molecules called Phosphoglycerate (PGA). The ATP will be converted back to ADP and NADPH will return to NADP+; these two molecules will be recycled in the light dependent reactions. The overall process is repeated and eventually the PGA will bind with other PGA molecules forming glucose. Not all

PGA form glucose though, some will be used as a recycled molecule in the process.

There we have it guys! Photosynthesis! All of this to release some oxygen, create its own food, produce ATP and overall survive. There is not a single step that can be missed in the whole process. How would we explain the conversion of PGA to glucose without the use of NADPH+ and ATP? All of these enzymes responsible for photosynthesis were once transcribed and translated into their function through gene regulation. I hope you guys see what I did there. From gene regulation to photosynthesis, there are so many more complex systems I cannot fit into one book.

I'm pretty sure the last time some of you guys learned about photosynthesis was in high school or as a required introductory course in college. Of course, we are learning it just for the grade, but this time I present it to you all to make a case for complexity. Even if you did not know a single thing about what was explained, just know there are thousands of other processes that are just as mind-boggling.

A Beautifully Orchestrated Death

The next topic shall be programmed cell death, more specifically apoptosis. Oh how macabre that sounds.

Well, yes! Your cells commit suicide. Again, this sounds dark, but the process is structured almost perfectly. This process occurs when mutations in DNA cease functions in a cell. Any cells activity that poses a threat to the organism must be neutralized. These cells are considered bad apples and will be removed by a phagocyte.

If apoptosis does not occur, tumors and cancers are formed. This explanation is a bit simple. I hope we didn't forget why we are here. Let's get started with this complex "you know what".

Apoptosis begins with splitting up a cell that is faulty or possesses damaged DNA beyond repair. The breakdown of the cell begins with a family of caspase enzymes (be aware I will use proteases and caspases interchangeably). These caspases hang out in the cytoplasm of the cell and are considered procaspases since they are not activated. Activation starts by signaling that occurs outside of the cell by T- lymphocytes (extrinsic). The lymphocytes come in contact with the surface of the cell and attach its FAS ligand to the cell's FAS receptors. This triggers intracellular events that start apoptosis. A FAS associated death domain called FAAD aids the process.

The word caspase is shortened for Cysteine Dependent Aspartate directed Protease. Caspases are protease enzymes responsible for cutting other enzymes. In the active sites of these enzymes is the amino acid cysteine. The activities of the caspases are dependent on the cysteine (I wonder what would happen if it were removed or had evolved from something else). These proteases (caspases) are designed to recognize the sequence DEVD [D=Aspartate E= Glutamate V=Valine]. The proteases then cut proteins at the site in which there is an aspartate linked to a glutamine, valine then another aspartate. This occurs multiple times between a series of other caspases in which it is called the caspase cascade.

The initiation of the caspase cascade then triggers a series of events that happen intracellular by an intrinsic pathway. Within the mitochondrial membrane are anti-apoptotic proteins. These proteins are BCL-2/BCL-x and BAX/BAK. In a healthy cell these proteins are bounded together blocking the process of apoptosis. When a cell is damaged BCL-2 and BCL-x are blocked. This allows BAX/BAK to embed themselves into the mitochondrial membrane. Cytochrome C is leaked out into the cytoplasm and bind to APAF-1 proteins that create the compounds necessary for the caspase cascade. This forms this parade of activation of other enzymes. For simplicity, I shall leave the explanation at that since it can become even more complex than this (oh the irony).

58   A Paradoxical Life

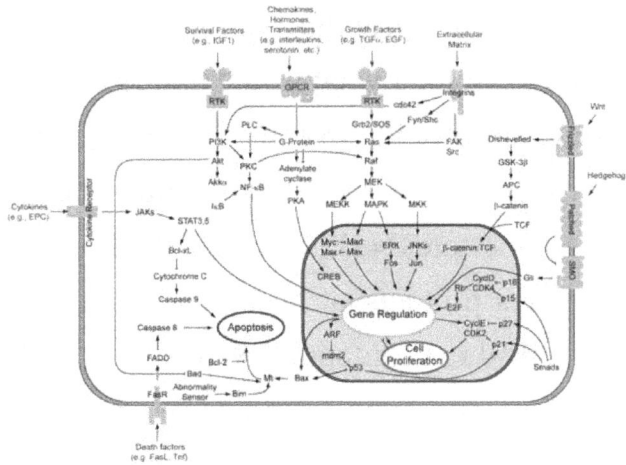

As shown in the figure above there is a plethora of caspases activated by a chain of other biochemical reactions. The parade of enzymes all results in the death of a cell. Who knew that death would be this hard!

Ubiquitin: A Somewhat Beautiful Death

There is one last complex biochemical process I want to sneak in here to really make the sale. P.S. I was writing the chapter about religion/spirituality and scrolled all the way back up to add this in. You guys are a spoiled bunch I swear! In addition to the caspase cascade and destruction of cells, proteins have a very complex process that destroys faulty proteins. The protein responsible for this action is called ubiquitin. This does not excite me as much as the process of the caspase cascade, but maybe it will excite someone else. I would like to give a fair warning before I continue; this is going to be complex!

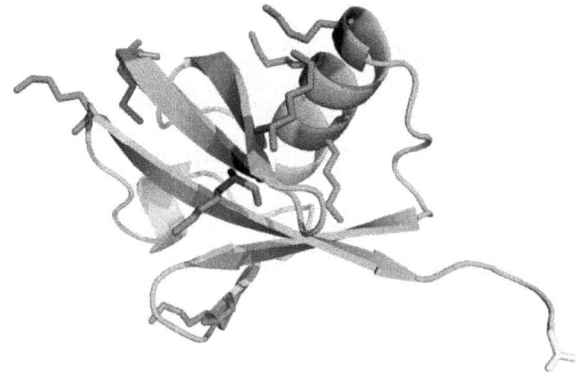

Ubiquitin is a small protein that is made up of 76 amino acids. It is present in eukaryotic cells and tags faulty proteins. Ubiquitin flags these proteins by attaching to a lysine side chain of proteins in a process called ubiquitination. More specifically, the c-terminus of ubiquitin attaches to the lysine epsilon amine of the targeted protein. Not only can ubiquitin attach to other proteins, but also it can attach to other ubiquitin forming a poly ubiquitin chain. One of the most common functional poly ubiquitin is by K48. The k48 protein is recognized by a proteasome and will cleave the k48 into a short peptide. The attachment of ubiquitin to a protein is catalyzed by a series of cascade enzymes called E1, E2 and E3. Here are some of the steps:

Step 1: The ubiquitin C-terminus is attached to the active site cysteine of the Ubiquitin activation enzyme (E1).

Step 2: The activated Ubiquitin then binds to the Ubiquitin conjugating enzyme (E2).

Step 3: The ubiquitin is transferred from the E2 to the lysine on the targeted/ faulty protein.

Step 4: Once the ubiquitin is attached to the target protein, more ubiquitin molecules are attached to that ubiquitin to prepare for degradation.

Step 5: After ubiquitination occurs via k48, the proteasome complex can find that target protein and break it up into its constituent amino acids.

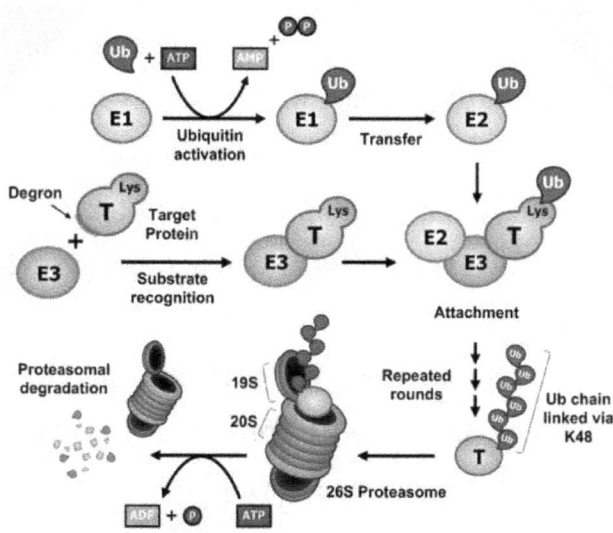

The overall process of post-translational modifications is programmed so well. We now know how proteasomes know which proteins to degrade.

These complex structures leave me in awe, and I hope it is having the same affect for you. It is also fascinating how similar it is to the processes mentioned before; all the proteins work in conjunction to one another. There are no useless parts of the system. I was fairly generous when explaining this process; there is actually much more to the process that could not be included. There is energetics, thermodynamics and mechanics of these molecules as well.

When we tie all of this together, photosynthesis, gene regulation, ubiquitin, DNA replication, the Lac Operon, and programmed cell death, what do we get? Yes, some complex stuff. They all share the trigger and switch method of operation. I would also like to point out that all of these processes are occurring at the same time and in large numbers. There are thousands of ubiquitin reactions going on at the same time. There are thousands of reactions involving the breakdown of lactose.

Through the use of specialized proteins, life for organisms continues efficiently. Biochemistry and molecular biology help scientists understand life in the best way possible. I am a bit biased here, but it makes the most sense. I am not talking about the evolution of these processes or the mechanics of the atoms that are found in these proteins' structures. At what point may we say something has evolved into another caspase or a membrane protein that allows atoms inside or out the cell? I introduced this section to really display in detail the uniqueness of molecular pathways. Some would say that humans are poorly designed, however at the molecular level we would be considered incredibly designed. I will touch on this topic much later as I digress. I believe the examples I have left is an adequate amount of information regarding the complexities of life. Are you convinced yet?

Part III: We Understand How It Works, Not How It Came To Be

Chapter 3:
Religion, Spirituality & The Unexplained

Welcome to those that stuck through or just skipped the last chapter. I promised not to keep that part of the book too dense, but how else would I be able to get my point across? How many head scratchers do I have so far? How many of you had to stop and look some of this stuff up before continuing on? If you did, do not be ashamed, that is a part of trying to understand the complexities of life in a molecular way. In this part, there won't be too much science talk. I would consider these chapters to be philosophical. I believe you guys will all feel the shift in the book with some of the upcoming topics that will be presented. We shall see things at more of a surface level; I consider this stuff you and a stranger can talk about. I am pretty sure strangers do not talk about 3-methyl 4-(1-methylethyl) 5-(1,1-dimethylpropyl) nonane on a bus ride home. They will talk more about politics, religion, conspiracies and our curiosities of where we came from.

I would like to also mention that there may be a lot of shaking of heads in agreement and disagreement. This is personally my favorite part of the book since it is so relatable to so many. I will have separate conversations with the Bio and chemistry geeks on my own time;

just email me. Those of you that are not inclined to the science talk, this will be right up your alley.

So, to please the crowd I will open up by briefly talking about religion and spirituality. This will create a good introduction into the next chapter talking about creationists vs. evolutionists. I see these chapters being the most controversial part since there really is never an in-between.

I've mentioned before that I am a believer in God, or some would call a Theist. A theist is an individual that believes in the existence of a God that had an influence on the creation of the universe. Though I am considered a theist, I still respect and understand the concepts of science. This chapter will help readers understand my stance on if religion makes a valid point for how we came to be. It should also show that though I am a believer in God, I still freely question things that would be practiced by someone who is a believer. Again, before anyone beats me up about this, there are a few points that I will make a rebuttal to. As you all know I try to beat you guys to the debatable topics. So here they are:

1) How could a scientist believe in God and science at the same time?

2) The bible says Earth is about 6,000 years old, yet science has evidence of the Earth being around 45 billion years old. Which is true?

3) How could religion explain the origin of life if at all?

4) Explanation for spiritual beings and is it just a figment of our imagination? Could this explain our origin?

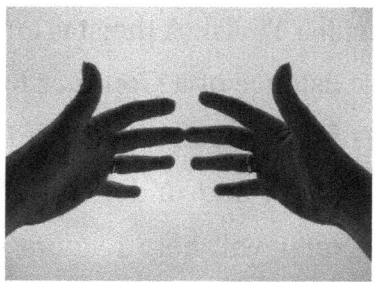

We typically relate the belief of God with the bible; however I am here to reverse this idea. Not sure if there are others that think the same, but God does not always have to be in reference to the bible. As mentioned before, I am a firm believer in a higher power. (1) The only difference between me and other theists is that I am a semi-theist. I believe in God, however sometimes I do doubt biblical stories, spiritual practices and religion (more doubt in religion than spirituality). I am a scientist that doesn't use the bible to argue the idea of intelligent design; therefore, this book has the small subtitle,

"A Biochemical Approach" and not "A Biblical Approach". I use biochemistry and molecular biology to explain the works of life on a more personal level. I practice some parts of spirituality and religion for a sense of purpose. Science does not explain the purpose of life; it only explains how it works. The next argument is science says that Earth is around 45 billion years old, and the bible says around 6,000. (2) With carbon dating we have fossil records that date back millions of years ago. As a semi-theist, I do not disagree with that evidence. So yes! I believe in an old Earth. I believe the timeline in which God created Earth dates back even before the Big Bang occurred (don't shoot me yet since I will go into more detail about this in chapter 4). (3) Religion cannot help explain the origin of life but can be used as a guide for life and distinguish from right and wrong. It creates purposefulness for life and combats our insecurities. Humans go nuts over the unknowing; in response to this, religion was created. (4) I am saving this one for the spirituality section (I cannot wait).

Now that I have defended myself for being a scientist that believes in God, I can move on to discussing religion and spirituality as subsections. These subsections will dive deeper with some examples that should leave you scratching your head after reading. Already a few chapters in and I am 100% sure the scratching is not my dandruff anymore. So, let's take a leap of faith into religion to help explain that the possibility of a God is likely.

Religion

It is pretty tough to define religion since there are so many aspects to it. In Rebecca L. Stein and Phillip L. Stein's book, "The Anthropology of Religion, Magic, and Witchcraft", they define religion as this:

> Religion is a concept constructed by the human mind that includes a particular set of human beliefs and practices. As a cultural construct it is strongly influenced by culture and by philosophical and theoretical backgrounds.

I use this quote to start with the idea that religion gives a set of rules for living life. This is all based off of a particular culture, hence why there is a spread of many different religious beliefs across the world. For this reason I do not use religion as a valid point for the possibility of there being a God. However, just having the construct in the human mind does. Let me explain. If we were to have a fresh restart on Earth with religion and spirituality never introduced to us, then would we ever believe in a higher power then? Hold on, let me even rephrase this again for some. If religion and spirituality were never introduced, would curiosity not lead us to find purpose and a higher power? Looks like I get to coin a term here; I call this, "Inevitable Human Curiosity". Before I define this term, I want you all to hold on to the point I made about resetting the Earth since it will be bought back up later. With inevitable human curiosity, we will try to figure out our origin no matter what. The seeking of this divine power will eventually lead to some sort of spiritual practices (not necessarily a religion) first. My case is that since humans have this curiosity in their subconscious, the presence of a deity must be possible. Even for those that are atheists in a new world without religion, I would see them succumb to some supernatural belief. It may not be God; it may be Satan. Worst case, it could be nothing. Regardless of what it is, there is some small search that every person cannot deny that they've had for a split second. Nobody could tell me they did not have that gut-feeling for a split second. The feeling where their existence was questioned.

Before I get into my idea of how we internally were born with inevitable human curiosity, let me introduce the other idea that some scientists may believe happened.

Early Hominids

One idea is that in the early hominoid days our ancestors underwent neuro-evolution. I want to introduce this idea first and then add my pizzaz into it. I hope you guys all trust me and have the patience for the foundation I am laying down for the future chapter; I want us to pinky promise again. Sweaty palms and a sweaty pinky. I believe by now I can be trusted. Okay, going back to some of the first pinkies (early hominid life). It is believed that evolution created the emergence of religion. Humans began to evolve and harness adoptive behavior and emotions. A popular example that developed was altruism. Many other species present altruism, but the way that humans act by altruism is more significant due to its turn to religion. In altruism, a species allows for another to benefit at its own expense. Now why would early hominids do something like that? The reason for this is all in multi-level selection. This is the idea that fitness benefits can accrue to individuals through group level effects. For example, a group of hunters will engage in co-operative hunting to catch bigger prey rather than hunting by themselves. Sure, this action may put an individual at greater risk, but there is a long-term benefit to this. The success for hunting a bigger prey is more likely and beneficial to the group. For the individual, bigger prey means more meat, even if it has to be shared. The multi-level selection approach requires social interaction. I want to really emphasis social interaction here since it is a big component of religion, so please hang with me here. Here is a timeline for the neuro development for these species:

1.Early hominids had very weak social ties and no group structure.

2. Eventually there was an increase in the development of neuro anatomy

3. Due to environmental pressure, social interactions were needed.

Through social interactions came the enhancement of primary emotions like happiness, sadness, fear and anger. From these emotions came the participation in rituals, group tradition and eventually religion. Curiosity spiked as emotion became more prevalent. Phenomenon that could not be explained led to explanations by social interactions and teachings of religious groups.

To reiterate here, this is only an idea presented by some evolutionary biologists. This idea may be true and does sort of fall in-line with my idea with a twist. I stand with the idea that there may have been evolutionary influence in the development of social interactions→spiritual actions→religion. Neuro evolution sounds more like an evolutionary biologist proposal, and it is. However, if the pressures were never there, I still feel that these social interactions would have developed from Inevitable Human Curiosity. Regardless of the scenario, these behaviors are internally programmed.

The Great Reset

I know some are tired of me bringing up this, "Great Reset" idea, but it becomes more relevant when we extrapolate this:

1. What would happen if we didn't reset the timeline for life once, but 1000x over?
2. Would the reset be the same each time?

My idea covers the answers to these questions. The timeline for life will most likely follow a similar timeline each time. The only discrepancies are the timelines possibly not being exact each time. Regardless of this, human curiosity, environmental pressures and evolution is inevitable. This is where I continue with presenting inevitable human curiosity. To start off I would like to ask a few questions for everyone to ponder on. For events in which something is unexplained by science, we turn to the supernatural as an answer. Would that be considered inevitable human curiosity (IHC)? I use IHC only for the origin of life and the connection to spirituality. No matter the number of times the universe recreates itself, there will be some newfound form of curiosity turned to spirituality. As stated before, I do not classify religion as a viable resource; it is only the mere subconscious thought in our minds of there being a higher power that I see viable. Just because I am a believer in God does not mean I die by the bible and its evidence. Just because I am a scientist does not mean I do not doubt scientific explanations. Some would say I am a fool for questioning Darwin and evolution theory, but please keep in mind this is just myself exercising doubt like any human being would.

The Fore

From here on out you will see me go back and forth on mysteries. Also keep in mind that the origin of life is considered an unexplainable event. The example I want to use to help show the difficulties of using religion to explain the origin of life is with a group of people called the Fore.

Before I continue with explaining the group, allow me to preemptively explain the significance of mentioning them. Let us remember that religion is centralized to group, culture and tradition. In this example we will see just how far the influence of religion goes in debunking the mystery that has made the Fore suffer for years. I would also like to highlight the idea of differences in how people interpret unexplainable events. We will see the scientific explanation and then the explanation from a group religious standpoint. Also, I would like for us to keep in mind of how this group developed these religious beliefs and if it falls under the idea of inevitable human curiosity.

The Fore are comprised of a small group that live in Papua, New Guinea. Attention of this group grew in the 1950's when Australian researchers found a very odd disease affecting the group. The mysterious illness was called Kuru (literally meaning trembling or laughing sickness). With the Fore being a very small population, the yearly death toll was enough to become a major problem in the region. Some of the symptoms they experienced were shaking, declined motor skills and the inability to eat. The symptoms eventually lead to a fatal death. The victims of the illness were mostly women and children; this was a pretty puzzling find. The original idea was that this might be due to an environmental cause. After further research, medical teams stationed in New Guinea concluded that the kuru came from a protein called a prion. These prions are transmissible spongiform encephalopathy proteins that are misfolded.

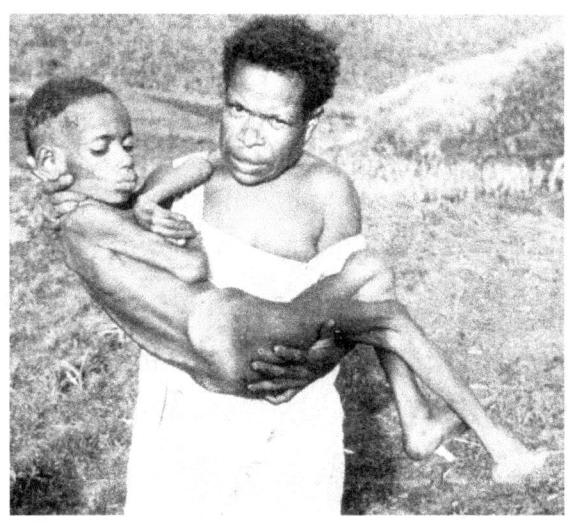

The next mystery was how could this vicious prion be carried to people of the Fore? This was all from the result of cannibalism. Yes, the Fore ate body parts of one another during funerals as part of their religious practices. They would have a family member that passed away and take them to a ceremonial field. They would cut the person into sections by body part and cook them. The cooking does destroy some bacteria and other microorganisms but does not destroy prions. After a few years of being in the bloodstream and headed to the brain of the victim, symptoms will then show. The reason children and women were more likely to have the disease is because they were more likely to eat the brains of the deceased. Their main choice would be the brain due to their social status in the Fore's social group.

It was difficult for scientists and medical groups to explain how prions were the cause of this all to the Fore. They did not take the scientific explanation and thought it must have been some sorcery or witchcraft. The fore believed that the prions were tiny spirits that cannot be seen that try to curse their people. To combat this sorcery, the Fore continued to practice rituals that would deface this wicked sorcerer. Though this seemed like an effective way to rid of evil, the solution would be to just stop eating each other. This is easier said than done.

You see, completely stopping cannibalism for the Fore would be against their religious practices. How could you tell someone or some people, "Hey guys stop what you're doing. It is not good for you"! You simply cannot; this group has been conditioned and shown for this to be a part of their culture and religious practices. There's disconnect between the actual science and religion to explain something as simple as: If you stop eating the dead, then you will not develop kuru. If you continue to eat the dead, then you will develop kuru. Again, it should be this simple, but it is not. There is clear evidence of why the disease started and a solution to it. Due to religious practices and ideas, individuals like the Fore say otherwise.

The last reason for religion's invalid reasoning for the origin of life is in its spread-out approach. Within those religions are many different interpretations. Religion is too broad of a topic to try to individually pick out what stories or books (Bible, Qur'an, Vedas etc.) make sense. There is over 4,000+ religions across the world. Some I would consider cult groups but let us leave that for another person to write a book about. Of the 4,000, there are 12 major religions. The first religion introduced to human life is Hinduism; this was the first major religion of India in the Indus Valley. Like in other religions, there is a sacred book that sets the paths to life and introduction of a divine power. The set of scriptures are found in the Vedas; it mentions the existence of Braham and other deities. The interesting thing about Hinduism is there is really no origin to it. It is considered timeless and had even existed before the writing in the Vedas. We have an unclear origin for where or when religion started; it is only a fair estimate. Following Hinduism was other major religions like Buddhism, Christianity and Judaism. The development of these occurred through the branching of previous religion and culture.

There are different books, ways of living and rules that exploded into human life. It is possible someone wrote the books in the way they see life should be. Over many centuries, translations and carryover of these religious practices could've been altered. Have you guys ever played telephone? How do we know which religion is correct with what is right or wrong? Was it wrong for the Fore to eat brains of the deceased when it caused disease amongst its people? Or it is right because that is what they were conditioned to. Right or wrong, the main point is that unexplained events are typically resolved with religious practice. The same is true for the unexplained events for where we came from. The conclusion is that we simply do not know. When choosing a religion, all we may do is learn about a religion and choose the best one that fits our culture, social status and well-being. Spirituality does not fall in the same realm of religion, at least the type of spirituality I talk about. For this reason, spirituality will be my anchor for the possibility of there being a higher power.

Religion Vs. Spirituality

I want to show the key differences between religion and spirituality. Religion is out of the conversation and debate. So all you "Great Debaters" against the origin of life through creationism can save it for some other person.

The reason for this section is to separate religion and spirituality. This must sound crazy to some since people think they go hand and hand. This is very true, however there are religious people who would go against some spiritual practices. Religion deals with an organization of beliefs and practices for a group of people. Spirituality is the individual practice where one seeks peace and purpose.

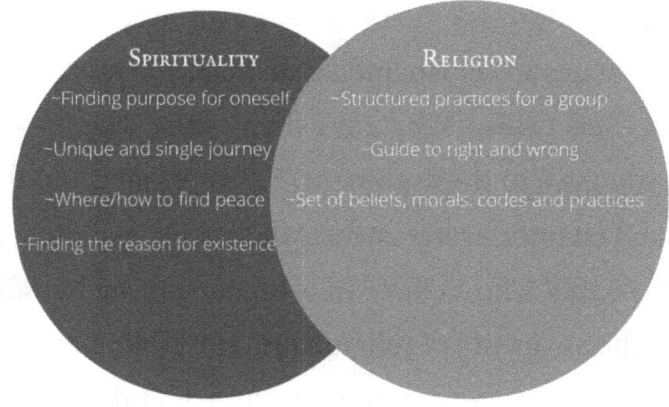

I set this up so perfect! This is where IHC comes in; I hope you guys didn't forget the term I promised to bring back up! Inevitable Human Curiosity is the reason for spirituality. This is a foundation for origin of life. Let's couple the ideas in previous chapters to recap:

Life is so complex there must be a creator→ Why is it so complex? →We were likely created by a higher power→ Due to our curiosity/unknowingness we inevitably/subconsciously seek our origin→Find spirituality/ purpose for life→ Someone couples all these ideas of their individual spiritual practices and spreads it to a group→ Religion. Are you guys with me so far? I hope the difference is made clear. I do not want to back my thesis up at religion. I want to start right at the curiosity.

Before I continue on, when have you ever seen someone use points and arguments not in support of his/her own thesis? Please give me a round of applause for not being so selfish and egotistical. I'm sure atheists didn't expect this. Why am I doing this? The idea is to beat you guys to the argument. I am setting grounds for the next chapter pretty early. Atheists immediately shut down people who believe in a higher power by talking about the absurdity of religion and the bible. I only want to use the complex systems in biology and a splash of spirituality or nature.

Spirituality

Now that religion and spirituality have settled their differences, we can finally give all the attention to spirituality. Again, be mindful that spirituality deals with a single person's journey to peace and purpose. Why do we seek this purpose? Humans have this natural, subconscious curiosity for purpose. In doing so, there are many ways a single person seeks this peace. We have rituals; certain music people listen to, a specific eating habit, lighting candles or prayers. I've asked previously if we were to have a reset on Earth, would we still be involved in spiritual practices? It is a simple, yes. Due to IHC, everyone would seek purpose and origin of life. People who are not spiritual today would seek some spirituality if the Earth were to reset. Not a single person just comes out of the womb and hits the "I'm not a believer button". It starts with being raised in a culture that believes in a deity. It is also pretty much second nature to believe in some out-of-reach power. Of course a person can then practice their free will and turn agnostic or atheist. Though a person is agnostic or atheist; they still seek an answer to our existence or have at least tried for a brief second. Their answers may not be spiritual or God-like, but nonetheless it is seeking something.

Let us think about this scenario for a second. We have an individual that does not believe in God, spirits, Allah, and the reality TV Gods (insert your God here) or whatever! Again, they didn't have that, "I'm a non-believer button" coming out of the wonder hole. Here is how it typically all starts out as. **FIRST STEP**: At some point in their lives, they had an epiphany saying, "I wonder where the hell we came from?" This is the individual practicing their inevitable human curiosity. **Possibility 1:** This individual then chooses faith forced by their family and culture. The individual can also grow up in a family that is atheist. **Possibility 2:** After a while this person believes a lot of this religion stuff is mumbo jumbo. They learn about spirituality and embark on their own journey rather than share it with an organized group. They are spiritual, but not religious. They may believe in God, but do not partake in the day-to-day practices of a Christian, Baptist or Catholic. It is likely that they believe in God, and only certain spirits. **Possibility 3:** The person has decided that the practices of religion are absurd and do not believe in anything; not a God, spirits, ghosts etc. **Possibility 4:** The individual can infer (notice how I didn't say believe here so I don't get the evolutionists here mad) life was created by a God, aliens, or through evolution.

I would consider all scenarios examples of an individual seeking spirituality. I say this because an individual is looking for an answer to the purpose of life; everyone has done this for even a split second. Why would we do this? That is because it is a subconscious action pre-designed into our human mind. We are the only species on Earth to be able to think this way. As I am writing this, I looked out of the window and tried to look at the ducks that fly across my balcony. Do these ducks think the way we do? Do they think about the origin of life? Well, I might as well try asking one myself. One second....

Waiting..........

Unfortunately, all I got was a quack back. Their thinking seems to be more programmed. The thoughts of these ducks go no further than their instinctual behavior. Compare that to the thoughts of a normal human or some crazy scientist like myself. What I'm trying to say here is that humans think out of the scope of any other species. I am highly confident neither the ducks nor the lizards by my balcony have any interest in the origin of life as we do. We have this undoubtable uniqueness about us that really goes over our heads. For us to be able to even think this way is a separate story of its own. Could this not be a result of pre-programming by someone or something? With this question I know some will immediately say, "God!" Well, what will my other people answer to this? If you are not ready to answer this question, then let's look at some more weird phenomenon that may make the question a bit easier to answer.

The Unexplained

Humans all have some sort of supernatural ideas whether we like it or not. When I say supernatural, it does not necessarily mean a God. This can mean anything beyond human reach in the physical world. So yes, this includes the ideas of ghosts, aliens, spirits and more. If a person can believe in aliens, time travel and multiverses that are not observable, then why can't the idea of intelligent design and be plausible?

I know this is a biochemistry-based book, but I want to introduce something that may have some of you guys really thinking. Interesting enough, this will give an introduction into the next chapter of a nice hypothesis that I believe. I will finally be giving astrophysicists, cosmologists and theoretical psychists some limelight. I mean, it wouldn't make sense for the cover of this book to be a photo of the cosmos without talking about it! I will only use this to support the complexities of life, so I promise to not get too caried away here.

Dark Matter & Dark Energy

Without further ado, I want to introduce a mind-boggling phenomenon called dark matter collision theory. In this theory, it is believed that not only do protons and electrons make up matter, but something called dark matter exists all over the universe. There is also dark energy that is responsible for the expansion of the galaxy. Both matters are head scratchers for scientists, especially dark matter. Dark matter is the glue that holds galaxies together and accounts for a much larger percentage of the matter in the universe. This wonderful matter is said to be running through your hands and body write now as I write this. It's an unseen particle that have psychists dedicating their lives to try to capture this thing in a jar. I hope your eyes grow as wide as mine when this next statement about dark matter is mentioned. That is, dark matter is not observable…… I hope this sounds very familiar! Psychists describe dark matter as something that we all know is there, but just can't detect it, a presence of some sort. This sounds a lot like when theists say something as broad as, "we know God is not observable, but he is here. We feel it in our bodies and our souls. There is the gut feeling".

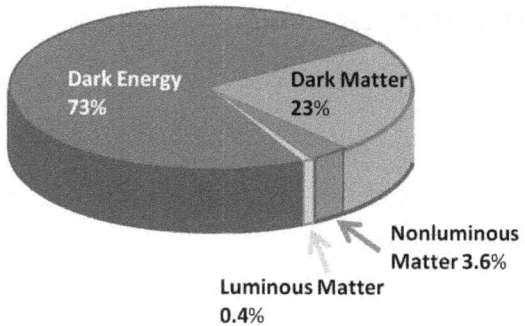

When we take an unexplained phenomenon like dark matter and compare it to the idea for the presence for a creator of the universe, what's the difference? It is all just guesses and hopes of finding a presence. Scientists want to literally catch dark matter and place it in a jar if they could. I know scientists that die by the gut feeling of proving the existence of dark matter. So why not have this same openness to others that believe in the presence of a higher power? We are all equally in the "I don't know" bubble. I hope astro and theoretical physicists are not offended when I say scientific theists are all in this category too. Do you know why we are in this category? This is because we aren't 100% sure of some of these things; they are mere theories and ideas. This is why I respect any other scientists and their ideas of the origin of life. As stated before, where I would like for us to evolve (hehe) in the acceptance of having an open mind and pushing egos to the side. When are we going to say something is unexplainable?

Chapter 4:
Creationists vs. Evolutionists

Finally my favorite part; it is the part that causes the most controversy. All that complex biochemistry is done and so is the spiritual talk. I hope you have all enjoyed some of the examples mentioned in those parts of the book. In this part we discuss the arguments between creationists and evolutionists. So let's put on our gloves and enter the boxing ring. Okay guys please don't do this. This whole topic is not to have some big brawl nor discuss who is right or wrong. My goal is to set a common ground between the two ideas. First, let me describe these groups of people first.

Let us start with the creationists. These guys are a bit quirky. For decades most evolutionists think of creationists as illogical thinkers and a bit nuts. I do not completely disagree with this statement either. I mean, the part about being a bit nuts should be very agreeable. This agreement does not mean I'm letting evolutionists off the hook either. The mind of a creationist is one that believes life was created by some divine power. This is the part I completely agree with. However, some of the points they make are usually biblical. Now this is where I disagree. I am pretty sure that you guys understand my stance by now; I just like to constantly reiterate my belief in God, but not to the extreme of using the bible as evidence. I would say the bible gives us a blueprint of how to live life by stories that were written long ago. So would I call myself a semi-creationist? Well there are different types of creationists in the group. Let's look at some of the three different types of creationists:

> 1. The individual that believes in God and refutes evolution, they strongly use the bible as a support to the origin of life, a way of life and the key to overall well-being. You cannot mention evolution for one second.
>
> 2. This individual believes in God, but still questions some biblical stories, evolution and the general conversation about the origin of life. They would consider themselves spiritual, but not fully committed to any religious group.
>
> 3. The conflicted bunch. These individuals are religious. They have a clear understanding of principles in science and see logic in some parts of the Theory of Evolution. However, they do believe in God and follow the bible.

Evolutionists are usually strong-minded individuals that understand science and believe the theory of evolution is the only thing that makes sense. They too are very weird when comparing them to creationists. Some of these evolutionists are people who hold high PhD degrees while others are ordinary people that have learned about some of Darwin's work. Lastly, we have the very toxic evolutionists that down anyone who dare to question the theory of evolution (there are toxic creationists too). Some individuals will quickly debate anything and are very closed-minded; they will immediately ridicule any person with a science degree that believes in God. It is pretty striking to speak with some of these individuals sometimes.

Dictionary Lesson 2

We usually go into a philosophical pit hole and are drowned with egotistical debates. These debates usually end in disrespectful and condescending jabs at other people's "credibility" in scientific knowledge. I digress here; I want to start getting into some of the arguments either presented directly to me or I have seen from others:

"I'm poorly designed, I wake up with back pain and have so many diseases, so I was not intelligently designed."

Again, these are statements typically seen by opposers of intelligent design. I am not including too many evolutionists in saying something like this. They would agree with a theist in saying we have a great design, but not the idea that it was designed by a creator. To reply to the statement mentioned above, I want to pose a question. Typically, imperfection in the construct of human life is used to discredit intelligent design. For this reason, we start with the question of, is intelligent design synonymous with perfection? The simple answer to this is, it is not.

I believe perfection is subjective like how art is. Some paintings and sculptures are pleasing or displeasing to some people's eyes. Look at some of the previous illustrations in this book. Would it be surprise to you that some of these illustrations were made by me? I am far from some expert artist, but I can say I did create some of these diagrams respectfully. Some of my friends and people who are close to me would probably say that it is perfect; the diagrams were drawn nicely. Others would say that my diagrams are lousy and lack any creativity. The point I want to make is that perfection can be perceived differently amongst each person.

First let's look at the definition of perfection. Perfection is defined as having all the required desirable elements to something. In the case of what I write, I can agree that humans have all the proper elements to satisfy life. Again, I say this in a general sense where if we consider the population on Earth, we can say that many people have all the required elements to function. Disease and back pain are only just caveats to the perfect system we have to sustain our lives. Sure, these things are not great, but not enough to discredit the incredible design and functionality of organism's lives. If you all need a refresher on how intelligently designed life is, you may want to check back in the molecular biology section.

When we say something is intelligently designed, it is recognizing the complexity for all of life. To completely toss the idea away for some mere hiccups in the design is not adequate. I had mentioned that perfection is subjective like art and this statement stands true no matter what. It may sound archaic, but I'm trying not to make this more difficult to understand.

As I look at some of the art in my apartment, I see perfection. There is a small chip in the corner of the frame and a missing clip that is used to hang one side of it up, yet it does not completely throw off its purpose and functionality. The painting still does the job fairly well. I look at it and it makes me happy. It still can hang despite a clip missing; it serves its function. When I look deeper into it, I value the creative mind of the person that painted it. Again, this example is really archaic and may receive some rebuttal like:

"This is a terrible example of intelligent design. Some people are born with conflictions. Your painting may have received these flaws over time. The comparison can't be made." Though this statement stands true, I still wanted to use the example to lay down a foundation. The idea is that despite any flaws in creation over time or at the moment of creation, there is an intelligent mind that had constructed it.

Dictionary Lesson 3

Okay here is the next dictionary lesson; I want to define intelligence now. We have a clear idea of what perfection is. Intelligence can be subjective as well. Intelligence is the ability to acquire knowledge and apply knowledge and skills. If we bring back the painting in my apartment, not only is it perfect to me, but I also appreciate the knowledge and skill applied to painting the piece. In the previous chapters when I discussed these complex systems, I did it for a reason.

The construct of the systems had to have a creator with extraordinary knowledge and skills. I apply this idea to the big bang theory, subatomic particles and the expansion of galaxies and universes to human life today. Maybe some of you aren't the artsy type and think that all art isn't that great. I want to bring in a more practical example as I have done throughout this book. Let us stop and think about electric vehicles. Now, I don't want to make this too complicated, but in your head think about how an electric vehicle works. I am sure plenty of you wouldn't even know where to start. None of you have even ever thought about how one works. The idea is that they are designed intelligently and are technical. It is not until the electric vehicles gives us problems or glitches where we say the person who made this thing is an idiot. Maybe the battery has faulty code, the window stops working or the headlights flicker before starting up the vehicle. It is agreed that all of these faults would be very annoying to us. We may kick the hood of the car and call it all types of names which I cannot name in this book. But does this mean that we discredit the intelligent design? Do we consider the creators to be non-intelligent? Well, for anyone saying, "yes" then you must have failed in creating an electric vehicle yourself and are jealous of the huge mind of the creators. I think it's safe to say that we are intelligently designed despite the individuals complaining about disease, back pain or some electric vehicle that has trouble driving itself sometimes.

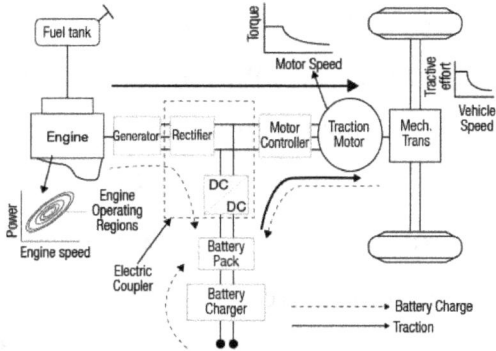

Something From Nothing

Another argument, or idea shall I say, is that the creation of the universe and life was spontaneous. An important thing to consider about creation is the idea that something cannot come from nothing. In the previous example about the electric vehicle, I explained its complex design and how it must have derived from a creator. In terms of the universe, the same is true. The universe could have not just appeared out of nowhere. The earliest idea of "nothingness" we have is right before the big bang and some funky matter. Before the big bang it was believed to be small clusters of dense matter in a steady state. At some random point there was a change in the energy in space leading to what is called, quantum fluctuation. Followed by this was the big bang, inflation and then expansion of the universe by dark energy (hello again). The whole idea here is that everything was created at a single point in time, a singularity.

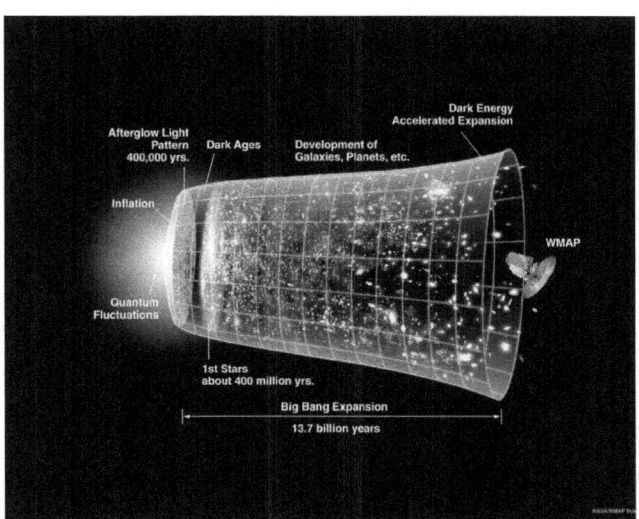

It has always been big bang theory & evolution vs. God. So how about we ask what was there before this dense ball of steady energy? Well, I can at least start by answering this with saying, we do not know. But to help a bit, we know that energy can only be transferred and not created nor destroyed. This statement is made in Albert Einstein's Law of Conservation. Okay, why am I bringing this up? Where's the head scratchers? If energy can only be transferred, then it could have not been created from nothing. Energy must have derived from some infinite entity. I hope I did not lose you here. Reread this over again before going to the hypothesis I've cooked up.

For your eyes only to read again: **"If energy can only be transferred, then it could have not been created from nothing. Energy must have derived from some infinite entity".**

Infinite Derivative Hypothesis

I have been foreshadowing this section for the longest. I was hoping you guys read a bit faster to get here (**rolls eyes**).

Okay by now I hope I've sold you on my idea by reading twice. Some readers are probably still folding their arms to this nonsense that I am proposing. Well, I still applaud you for sticking it through this far. I believe this will be a strong win for both camps here. Up until this point the idea was only to combat some of the problems with evolution theory and opposition to creation theory. However, with this idea we will see where evolution and creation may be able to merge. First, let us look at the next argument that is had: "God is not observable; we have more observable evidence from fossil records and plenty of backing for The Theory of Evolution. There is no evidence for intelligent design. It is just an argument, barely a hypothesis."

I am not going to hold anyone's hand and go over what the scientific method is. A hypothesis must be testable and falsifiable. I can agree with the argument of intelligent design being a tough hypothesis to support. However, it is easy to be open to the idea when you think about the holes that are in some other hypotheses presented to us.

Overall, the statement does hold true. I just want us to have an open mind before we start this argument. Let me start it off by bringing back that good ol' dark matter thing I was talking about briefly. We have astrophysicists that study dark matter and cannot see nor actually capture the thing. Though they say, "we just know it's there, we just know". So, when statements like this are said, is it not too bad to say the same for God or a Creator? The argument we have seen against God creating the universe is that the law of physics can make nothing out of something; God cannot. Of course, this depends on what you define as nothing. For simplicity, I mean this in a sense of there being particles present in quanta. I am far from an expert astrophysicist, but I am sure there is never really anything that is, "nothing".
As said before, the theory of evolution may be true. But even if we go back before this, we know that science says that life started with the big bang. Something had to have created the particles that collided and the particles before that and even before that. Sure, there are endless

theories that were presented by many scientists, but the fair answer is that we do not know. Is that such a scary thing to say? If we continue to question where something had started from, we go down a path of infinite derivatives from one thing to the next. I am sure this is something a creationist and evolutionist can agree on. Let me run this by you guys again! From both sides, if we were to agree that evolution is true and the big bang holds true as well, what made the components before the big bang (I shall bring this back up later). Did we have a creator that only created the elements first and then future date his/her creation through building blocks (which would be elements)? It is kind of like someone sitting at a desk and creating the smaller parts to their creation and letting it free form (through evolution). For those that have more of an appetite for visuals, here is what I am talking about:

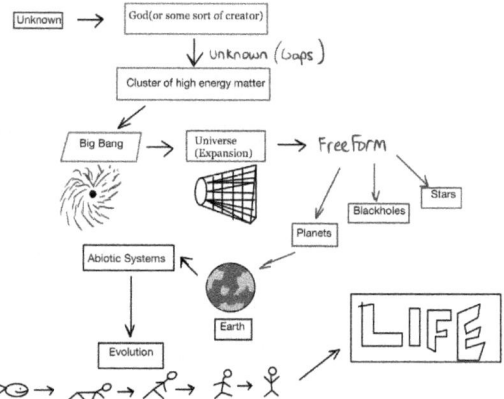

I am pretty sure Darwin lovers are no longer folding their arms anymore, but probably threw the book to the wall already. Others are reading this with an open mind as I am. With all the theories shared by the science community, this tree gives a firm idea of what I see possible for the origin of life. Finally a scenario of, "if Evolution Theory is correct". Keep note of the possible timeline here; this is considering creation theory and evolution theory. I swear it doesn't get any better than this. It starts with God or some sort of creator creating the infinitely dense cluster of matter or whatever was believed to be before the big bang. After this, the big bang occurred and began the expansion of the universe. Star and galaxies continue to expand and eventually form planets like Earth. We all know of Earth being a pool of heat and then later having formed gases and abiotic systems. From there, systems had evolved into life. This is the part where the higher power's creation was in free from. As stated before, it may have been some future dated design made from a creator into this complex interconnection of life forms.

    The way I like to think of it, we can compare the creation of life as Lego pieces. In each Lego set there are a certain number of pieces needed to create a structure. The structure can be a warrior, spaceship, a city or a sculpture. Please be imaginative with me here. Let's say we were to throw pieces from the warrior set, the spaceship set and a few pieces from the building set into a staging area. It is quite obvious all these pieces are different. However, these pieces eventually fit one another. The final product is a City Space Warrior Sculpture that's created. The creation of the final product may have been random, but the pieces to create it were probably not. God represents the creator, and the Lego pieces represent the atoms, matter or whatever was there before big bang. The pieces were strategically chosen to create what we know today but may have randomly been assorted through evolution. Though this example only explains the timeline of God to Big Bang and then to human life, before God is what the biggest mysterious is.

There seems to be an infinite timeline for the creation of life before the big bang and certainly gaps in between what we believe we know. At the molecular level I've expressed enzymes that serve as precursors to other enzymes and question the function of some of the enzymes before. I had asked what components made the big bang and the matter even before that. To combat this question, I want to introduce to you guys of something I call the infinite derivative hypothesis. Be sure to not confuse this with the derivative of infinity function in calculus. This is a term to define the idea that matter and creation of one thing to another derives infinitely.

> ∞ <--*Unknown*< ---*Unknown*<--***God (or creator for the sensitive)***→Cluster of Matter→ Big Bang→Universe→Galaxies→Planets→Earth→Evolution→ Life

If we continue to ask where one thing had derived from, we would probably be in for a long conversation about the unknown. Though I am a believer in a creator, I am curious about his/her origin. It is a continuous debate that have all scientists puzzled. For now, we can accept ideas that can be prospects for finding the unknown. There is no observable evidence for God nor is there an explanation for the unknowns before God.

Things are way different when God and the bible are not used to back the mysterious questions about the origin of life. I believe I deserve a round of applause for this consistency. I don't combat each debate with, "we don't know, so God!' I haven't referenced the bible in any support for my idea. I prefer to be more rational in this process.

Part V: One Base Two Boots

Chapter 5:
The Idea of Collective Origin

In this chapter I want to piece everything together for those that still probably have their boot on one side still. I will also sprinkle in the idea of aliens and hybrid species very briefly since I think it's worth talking about. I really changed my mind at this point and decided why not. I think that taking religion out of this was critical; there was not a single bible thrown at any argument against evolution theory. What I did was integrate God, the universe, spirituality and science into one home. I would hope we have of our boots on the same base by now. I also hope many of you noticed what I did in the last chapter. Not only does the infinite derivative hypothesis support the idea of evolution, but it also supports the idea of creation and any other ideas that can be thrown into the mix. I made a pinpoint for God being the creator for the start of the universe and anything before that. I am not saying evolution is not possible, nor saying that creation is impossible. Instead, the two merge into a grand scheme that I call a Collective Origin

For the last time let's create some bullet points for what the Collective Origin accounts for. As I bring up other subsections, they will fall in line with my idea. In summary, the Collective Origin summarizes all ideas presented into the book. Collective Origin accounts for:

1. Unexplained Events that are not only backed by science (i.e., spiritual events, miracles etc.), but spirituality and deeper forces inside of ourselves.
2. Backs evolution: accepts the idea that species have evolved through a certain timeline. Ideally, pre-designed pieces that free form into the creation of life as we know today.
3. Pre-designed pieces help build the universe and eventually life. This led to inevitable curiosity.
4. Curiosity leads to finding purpose in life. We turn to spirituality then form religion.
5. We understand that life is complex and one of the best explanations may be a creator. Whether that creator is God, aliens or the Reality TV Gods (i.e. whatever you believe in).

Collective Origin

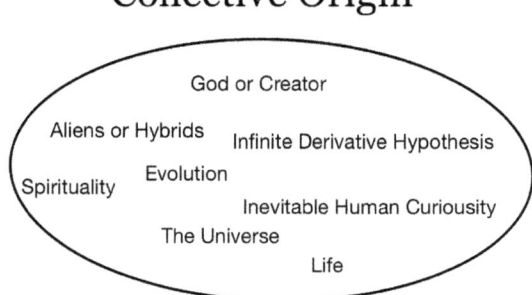

Aliens & Hybrids

I really didn't want to go down this road. After careful consideration, I've decided to add the topic. I cannot neglect the fact that it is an idea for the origin of life. I want to respectfully open the discussion for people that believe our origin are from outer galaxy aliens. Before I go into how it may be integrated into collective origin, let me give a history lesson and explain how others think we derived from aliens.

I will split this section into subsections for why this is a thing. After this, I promise there will be room for aliens in collective origin. And per usual, here is a chart for you spoiled visual people:

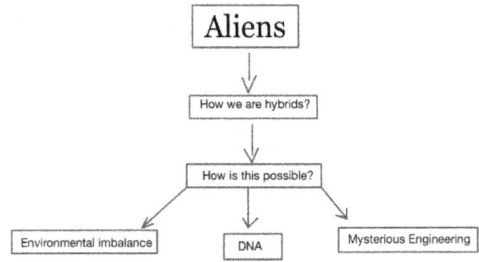

We can see this chart where I set the subsection into the reasons aliens may be a potential idea for the origin of life. A lot of these reasons have already been discussed in previous chapters but will really glue into one general idea.

A Brief History of Aliens

It is really difficult to give a history lesson about aliens. Well, here I am being that fool to give you what you guys want. I would also like for you all not to get your hopes up with some history textbooks with crazy dates. The reason for this is because there have been reported sightings before humans were even a thing. When humans were around, they probably viewed the strange sightings of UFO's and aliens as spiritual encounters. The typical physical appearance is usually a huge cranium and very slender body. It is somewhat like a human, but very peculiar.

Aliens really fit in the same category as someone who may believe in God or swears the presence of dark matter. I say this because, in all scenarios we say, "we know it's true; I saw it". For example, if we were to meet someone that says they've been abducted by aliens we would probably look at them crazy. Tell a pure evolutionist (one who may not agree with my collective origin and infinite derivative theory) or atheists that you believe God created everything, they would look at you crazy. The same is true for that guy you meet in the grocery store that randomly stops you and says that he's seen God twice in his life. He claims that he knew what he saw and that he has a gift for seeing divine powers. I know that a fair majority of people would probably think this guy is crazy too.

People who say they've seen aliens can fit in the same categories, so they deserve a fair shot. However, the whole alien conversation is a bit shaky. The history of aliens has been filled with hoaxes, lies, not much evidence and a ton of secrets. Though the history is shaky, I still want to open the discussion.

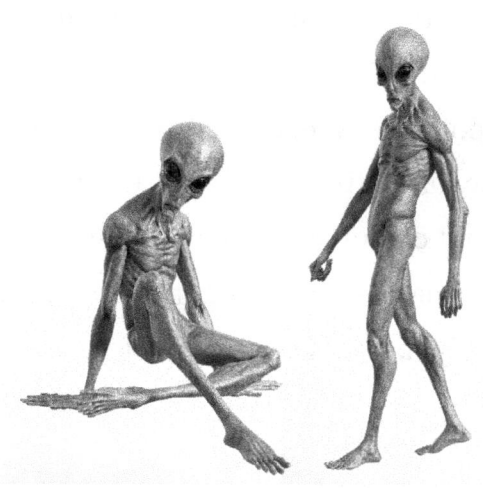

How Are We Hybrids?

This is the only example I will use to support the idea that we may have derived from aliens as a part of collective origin. The whole human/alien hybrid thing really stems from unknown phenomenon. Rather than pointing to the supernatural, scientists and others quickly draw their pointer to... Aliens! Pointing the finger to God or some deity we can't see seems illogical. I guess pointing the finger at aliens is viewed to be more logical since it is more scientifically sound. To keep things clear, aliens can be any type of extraterrestrial being outside of our planet, solar system or galaxy.

Yes, this includes those alien lizard people we hear about on backend forums. Again, I classify all these as assumptions to not knowing our origin. It is very difficult for the human mind to extrapolate phenomenon that is not so obvious. We can give aliens the same respect we do to creationism and evolution. To be fair, creationism does not always mean that we were created by God. This is why often times you see I will write, "some creator". I am not claiming to believe in aliens, rather, I am allowing for the idea to be a part of the discussion and collective origin. You see! Everybody is welcome to the party!

How Is This Possible?

Okay, how in the world is it possible for us to be derivatives of some weird alien species? Well, this is where I will have subsections with some of the most common theories you and some random guy at a bus stop talk about. Disclaimer: Do not talk to strangers unless they want to talk to you about a having a formulation that extends the telomeres in chromosomes or something.

Environmental Imbalance

I remember talking about those ducks that I see every day by my balcony. Well, if you need a reminder then I have no problem giving a recap of my conversation with them about the origin of life.......... Quack! Quack! Yep, that is all I got! I bring this up because I like to compare humans to every other species on this planet. We are the only species that are this complex in the way we think and live. Other species seem to have more of a balance with the environment. Humans, however, have a far less balance with the environment. I can blame this on the current technology we have, but in general our minds are way past dependency on nature. One of these lizards by my balcony do not have the mental capacity to create an electric vehicle or fly a spaceship to the moon. Maybe one day they will be able to do so when they evolve, but is that really what it is?

    I usually like to spend time opening conversation with strangers about mysteries, unexplained events and more specifically the origin of life The one time I spoke with someone who wasn't a stranger was at my day job about this crazy origin stuff. I remember sitting in the breakroom eating some pizza that the company would get us every Friday. Now, it wasn't typical for me to sit and eat in the breakroom, but for whatever reason I did. Of course, there were a parade of different conversations, but then aliens came about. I always knew humans were way different from any other being on Earth. What really struck me was when my co-worker told me to really think about it.... why are humans so unfitting to nature? With how damaging humans are to the environment it seems like we do not belong. We pollute the earth, damage communities and populations of other communities. This is all because of our unique mind and intelligence. We are the most advanced species on the Earth when you compare us to a rodent, a dog and even to great apes. We talk about climate change, a dying earth and even the exploration of a new planet to live on when things go left. Only WE think about these things. There is clearly an imbalance with humans and planet Earth. I am not saying that the Earth is a terrible place for us to live, but it isn't all so perfect. Humans are like a puzzle piece that sort of fits, not perfect, but just good enough to make the beautiful picture.

Mysterious Engineering?

The breakroom talk was followed by the discussion of mysterious landmarks that were built long ago. The most popular mysterious landmark are the pyramids of Egypt. We always have wondered how was it possible that these groups of people were able to construct something so massive. Again, we have a scenario where, if something cannot be understood, it is either by the powers of God or aliens. In this scenario however, aliens have won the vote. This vote assumes that the Egyptians may have not have the proper technology to move a two-and-a-half-ton block. Is it safe to say that maybe they were very smart and were capable? Well, you see, this does not work for most scientists. If we cannot replicate something like this modern-day, then there must have been some type of extraterrestrial help. The help from aliens is prominent since they were the creators of human beings. It is said that aliens visit Earth periodically and typically help us out. Again, these are just ideas that are formed from conversation and some theories made by some individuals.

I've mentioned before that our human DNA had been engineered by God or a creator. I hope you continue to see the verbiage used here. I have talked about the unique and complex construct of DNA in the beginning of the book. I want for everyone reading to at least stop and think about what DNA is. We can stretch the human chromosome and see all of the special genes that make up who we are. Outside of being God's creation or a result of evolution, our DNA may have been created by some extraterrestrial.

> ∞ <--Unknown< ---Unknown<--God (or creator for the sensitive)→Cluster of Matter→ Big Bang→Universe→Galaxies→Planets→Earth→Evolution→ Life

Here we go again! I want to bring this chart back to really drive this idea home. I really want to highlight this section of the chart:

∞<--Unknown<--Unknown<--God(or creator for the sensitive)

This is where aliens fit into the collective idea. I loosely use the word creator to allow everyone's imagination to be considered or what they deem to be the designer. Collective Origin allows for a more open idea to the wide range of philosophical views for the origin of life. Could I have been any more neutral? I would like for you guys to throw any other idea at the collective origin bubble. No matter what ridiculous idea is added, it can fall in the bubble. God, Alien-reptiles, Zeus, or Darwin. All can be considered for creators. Regardless of who the creator is, the complexity of human life still remains true. The unknown of where these creators came from is infinite. In all instances we have an infinite derivative of unknowns.

We say the universe expands infinitely and is mostly portrayed as a cone expanding in one direction. However, could this expansion just have been a sub expansion of a greater expansion? It is quite possible that we are just a small speck in the overall matrix. I want to be very specific here; when I mention the matrix, I am not talking about a simulation. I am saying this to push back on the infinite derivative theory and the collective origin. Let us take a look at these beautifully drawn diagrams:

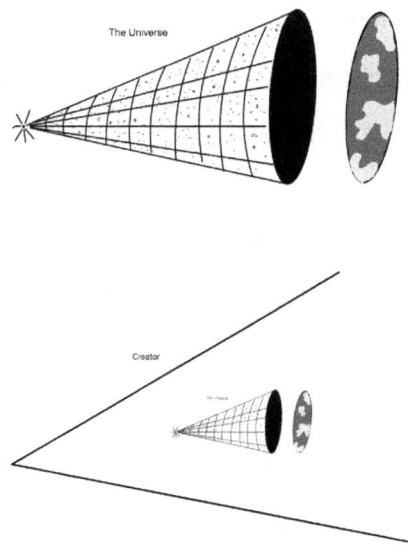

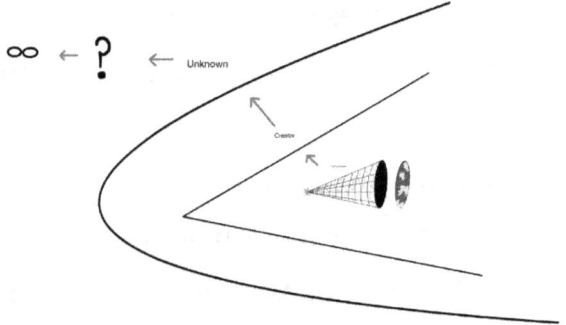

I'm excited if you still are tuned in towards this part of the book. I really want the diagram presented to be the final bang (no pun intended). It really gives an idea for what may have been before the universe. In no way shape or form am I claiming for this idea to be correct, but it does create an open discussion.

    The mysterious engineering (MystE) does not just stop at life on planet Earth. MystE spans to the very moment the universe was created. What is the overall idea here? I remember in the earlier chapter mentioning how can we have all the "-ist" happy and on the same boat. Well, that is exactly what the idea of collective origin has done. It incorporates the ideas of theists, Darwin fanboys and semi-theists; oh, and the reptile-alien believers. What we have here is an open-ended idea for what the origin of life may be. Again, I compare this to one of my childhood toys, Legos. You are given the pieces to some sort of object. You can stick to the instruction book to make the figure, or you are free to piece it together into what you believe is your work of art. I 've mentioned the possibility of the creation of life and the universe to be in free-form. I believe that the idea of collective origin does the exact same thing. All what are shown in the diagrams may be true, but maybe you do not believe that the creator was God or aliens. You are free to believe anything. Just know that whatever that belief is, it will always satisfy and fall under complex creativity of some creator.

Part VI: Close The Book

Chapter 6:

PSSM End

(A Philosophical End, A Scientific End, A Spiritual End and My End)

A Philosophical End

I guess this is the end guys, it has been an amazing journey for us all. Again, if you've made it this far you deserve a pat on the back. This will be the closer chapter if you haven't realized just yet. But wait! Don't close the book yet! I still want to revisit the main idea of why this piece was written. I also want to inform you that this chapter will be a bit more personal. This section of the book will share in detail more reasons for writing this book. I will organize this chapter with a topic followed by the word "End". For example, it starts off with' "A Philosophical End" followed by "A Scientific End" and so on and so forth.

At this point you probably know more about the infinite derivative hypothesis, collective origin and inevitable human curiosity than the author of this book(me). Most people skip the first few pages of books so I wouldn't be surprised if some delinquents skipped my introduction. For the naughty people that did skip, I can sneak in something that I said in the note to family, friends and my colleagues. That note states the confidence it took to start writing a somewhat controversial and philosophical book. As I write this, I am someone that does not hold a fancy master's degree or PhD. Even if I held either one, that would not hold me to a higher standard than anyone else. So how does this guy with a bachelor's degree in Biological Sciences even have the authority to share his philosophical thoughts.? The answer to this is that there is no authority. I am a curious scientist, matter of fact, a person with a fascination for the origin of life. In general, sharing scientific ideas and discussion can be quite scary. I created this book because I had a compilation of ideas that have always wanted to escape my mind. But the roadblock for me was mixing this with biological principles. As I write, the roadblock is not there anymore. I started to compile these ideas for the origin of life by expressing biological concepts that are deemed complex (I believe all are complex, but I am not trying to make this a textbook). The main goal was to bring together ideas from the scientific world, spiritual world and the weird, sort of pseudo-science world. I know this would sound impossible to some of the most profound scholars. But then again, what really is impossible? Yes! That is one of the main ideas here. That is, anything is possible because we really do not know much about our origin. For the fifth time maybe, I do not make a crazy claim that I know my philosophy is more correct than the next person. There could be no infinite derivatives or collective origins. For all we know, we could've been created by some unknown world ruled by cats. The universe and everything could just be a figment of our imagination. Then what is our imagination?

I've always questioned phenomenon in the universe and life. A stronger interest in philosophy came from taking a class during my undergrad course. Is that a surprise? I choose this class as an elective since I thought it would really go well with my knowledge in biology.

I remember sitting in the very first lecture and having my professor walk in. He was a medium-sized gentleman who came in with flip-flops, a scarf and a strong English accent. A very peculiar dress-style for a professor if you ask me. Not only was he different from the style of his clothes but, his lectures too. I was used to the traditional setup, the PowerPoint presentation and a laptop setup. For confidentiality and some sort of personable affect, we will call the professor Dr. G. He would walk into the class empty-handed with only his mind. He started going over the syllabus briefly. After this, he cracked open him marker and started drawing on the board. He started with asking, "what is this?" It was poorly drawn, but the whole class knew it was a chair. Everyone looked confused and so was I. I started to think, okay this is going to be an easy class. The guy is drawing diagrams for us and asking what we see. As we are in all agreeance that what the drew was indeed a chair, he proceeds to ask us, "How do we know that's a chair? What defines a chair? The English language calls it a chair, but it could be called a coconut for all we care." When I heard all these questions being asked, I really thought about it. I've never owned a rabbit, but I'd never back down from going into a rabbit hole here. After, he asked all these questions that is exactly what happened for the remainder of the 2-hour lecture. We went from talking about chairs to language, linguistics and then politics. All derived from the question of what is a chair?

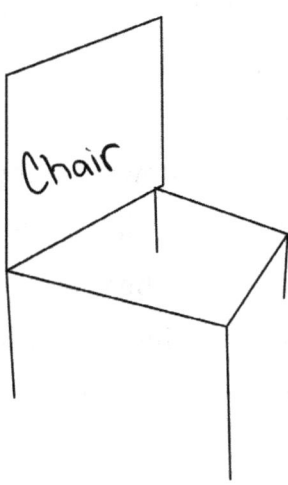

Do you all see what I am getting at here? The idea is that we can break the tradition of keeping science and other studies separate from one another. Also, we can come into any situation with a free mind like Dr. G did. The last point to take from the example and the most important, is that questions never stop regardless of what the topic is. The same is true for the origin of life. What is a chair? It is only a chair because we call it that. What is life and its origin? Do we just stop at evolution and God because that's what we were taught?

A Scientific End

Before philosophy, science had and will always have my heart. I've enjoyed the subject and every aspect of it since I was a child. I hope there are no second offenders that skipped the "About The Author" page. If so, let us go down this road again briefly. I pursued my education in Biology and have since then expanded to other general sciences to help write this book. As mentioned before, I was a bit reluctant to create a book centered around science and the origin of life. My inspiration for science really heightened when I took my undergrad course in Biological Principles. Again, I thank my university for some of the weird, quirky and eccentric professors that were presented to me. I started to learn more about genetics, sexual reproduction and then a bit of biodiversity. Each course that I took just left me with more questions.

Like the experience I had with Dr. G, I had another professor that helped shape my fascination. It started in my introductory biology course. We can call this professor Dr. B. This professor was a very interest character that would have loud outburst. I wouldn't blame him since his love for science was so great; one day I aspire to be just like him. Each lecture that I sat in I wanted more, more and more. The typical strategy for this would be to work in one of the professor's labs or take one of their special courses that had certain criteria during registration. This approach may have been easy for some, but very difficult for me. Let's just say I wasn't the best when it came to my GPA and other fancy accolades some student had. All I could do is admire the knowledge I gained. If we fast forward to the end of the biological principles class, I remember it ending with the missing link and origin of our species. I still have the image of the PowerPoint slide in my head. And like that, my mind has been left curious again.

When biology was coupled with philosophy, my mind had an inflationary period of wanting to know more. I wanted to sprinkle some of my ideas on this topic any chance I could. Today as I write this, the ice cream is topped with plenty of sprinkles and confidence. I want to us all to be able to sprinkle our ideas as scientists. Hell! We can all open our own ice cream shop at one point after that! The idea is that science and the passion to forward the thinking of the community drove me to writing this. What thoughts do you all have about the scientific community? Could we expand the thoughts of science into other avenues?

A Spiritual End

No this isn't about ghost, demons or some crazy spooky spirits. With science and philosophy, I never thought that spirituality would be in my vocabulary. I say this because it is typically absurd to accept spirituality as a valid explanation for the natural processes of life and its origin. Second, spirituality is usually looked down upon in a group of individuals who come from a religious background. Thirdly, spirituality can be very subjective. When one person mentions spirituality, it can be about prayer to a God, while another person can be talking about lighting candles and praying to spirits that are common in their culture and outside of the realm of Christianity.

I started to accept spirituality as a way for me to find comfort and peace. Each person has their own unique spiritual journey; spirituality is personal. My journey felt like it can be integrated into my everyday way of life. So, what do we have? A scientist/philosopher who is spiritual? Yes, that is exactly what you get. I like to speak for the individuals that feel conflicted about not knowing where they stand with science and the spiritual world. Are you on the same page as I am? Do you agree that scientists can be scientists and have spiritual a spiritual journey? I usually leave these open questions at the end for the reader to answer, but I have it this time! The answer is you can be a scientist and have a strong belief in spirituality. Science helps explain things in the most logical way. Some may say anything besides sciences is illogical, but we must remember that there are no rules. Like Dr. G asked us, "How do we know it's a chair?" Well, I have the same question. How do we know a scientist's ideas are the most logical? How do we know we're right?

My End

Philosophy, Science, Spiritualty and Me (PSSM). It saddens me that this is the end of, "A Paradoxical Life: Where Did We Come From?". All the ideas compiled to make this book were around PSSM. Even though this is the end of my talk about this topic, there will be more mind-boggling adventures to cover. The favor I ask of all readers is to expand your ideas like I have done in these books. Hmmm... I said, "these books". Yes, this is me hinting towards more to come. Like the universe being infinite, I plan on scientific and philosophical ideas being the same. I like to see what ideas you can bring to the table with confidence. Just remember that your ideas are never wrong (only if it's about a flat Earth).

Life is a paradox, so what is, may not necessarily be what we think it is. Or what you may not think it is, is really what it is. Yes, please read it again. As a matter a fact, read it three times more. Now I can ask my final question. What are your ideas in this life's paradox?

Write Your Idea Here:

END

Appendix

The purpose of this section is to give a little more bang for your buck. It was very hard writing my last few words as I ended the book for the last chapter. This is a good way for me to not let go of the book just yet! It's a win-win situation for both of us. I have said before that it would probably be best to brush up on some terminology, so readers are not too lost with some of the scientific terms. So, I would like for this to be treated as an extra for some information. Look at some of the supplemental information provided and definitions that follow.

Metabolic Pathways

Metabolic pathways are interconnections of enzymes that are catalyzed in a sequence to produce a final product. The enzymes used are usually recycled to continuously form intermediate and final products. Metabolic pathways are essential in all functions of life. They help build, breakdown and block formations of major molecules. Remember, these major molecules are used for life sustaining processes like glycolysis and photosynthesis. The 3 most common types of metabolic pathways are:

1. Catabolic Pathways
2. Anabolic Pathways

3. Amphibolic Pathways

Catabolic Pathways- The breakdown of molecules into simpler ones that will be used in another pathway to make macromolecules. For example, food has macromolecules that get broken down by our saliva to form smaller molecules that will be used as storage for later use. The other result can be the release of energy.

Anabolic Pathways- The synthesis of larger molecules from smaller molecules. An example is in photosynthesis. A glucose (larger molecule) molecule is built from carbon dioxide (smaller molecule).

Amphibolic Pathways- This is a biochemical process that includes both catabolic and anabolic pathways. An example is in the Krebs Cycle.

Charles Darwin's Theory of Evolution

The theory of evolution was proposed by English naturalist, Charles Darwin. In the 1850's Darwin wrote his most popular book, "On the Origin of Species". He documents his observations from a journey he took to the Galapagos' Islands off the coast of Ecuador. Darwin's main observation was that species evolve from a common ancestor. What Darwin had found were species that were identical, but still different. These species must have some sort of connection. He started by observing finches that were well-fit for their environment. Though some finches were different, they fit well. The characteristics Darwin looked at were the finch's beak size. Some were large, thin and others small. The reason for the variation in beak size attributes to the adaptations that make that specific species fit for its environment.

With adaptations influenced by the environment, Darwin proposed that these species can change over time too. The species he observed were believed to derive from a common ancestor. Over a long period of time, these species branched off repeatedly and are separated by splits from one species to the next. With each split was what he called, "descent with modification" that forms into a family tree.

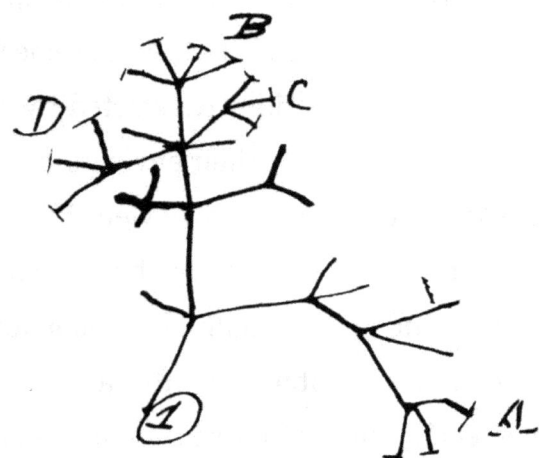

Another huge aspect of Darwin's Theory of Evolution is natural selection. Not only do organisms evolve, but populations do as well. The main concepts of natural selection are:

1. Traits are heritable
2. More offspring are produced to survive
3. Offspring's heritable traits vary

With these observations, Darwin concluded that some species would inherit traits to help make themselves better fit for their environment for survival. To continue this survival, individuals reproduce in hopes to pass these traits to their offspring. Since these heritable traits

are beneficial to survival it makes sense to want those traits to be more common across the population. Over generations, these traits become more prevalent
and continuously pass on and evolve as environments change.

What Is DNA?

DNA is an acronym for deoxyribose nucleic acid and are found in the nucleus of cells. DNA is an essential component to all of life. It is a set of instructions used for how an organism looks, functions and react to environments. DNA allows for reproduction between organisms and the spread of traits to offspring. DNA is found in viruses, humans and animals. This special coded instruction is made up of a long strip of nitrogenous bases: cytosine, adenine, guanine and thymine. In addition to these nitrogenous bases is a sugar phosphate backbone and phosphate group.

Definitions

Alcohol Dehydrogenase: A zinc enzyme that catalyzes the primary and secondary alcohols of aldehydes or ketones. ADH breaks down alcohol into acetaldehyde.

Apoptosis: A controlled part of an organism's growth which results in programmed death of cells.

Atheist: An individual that does not believe in any deity (i.e., a God or gods).

Amazon: A global distribution/ E-Commerce company that provides many products to its consumers.

Amphibolic Pathways: A biochemical process that includes both catabolic and anabolic pathways.

Anabolic Pathways: The synthesis of larger molecules from smaller molecules.

Andromeda: A galaxy located north of the celestial equator and about 2.5 million light years away. It is named after the daughter of Cassiopeia, the Queen of Aethiopa in Greek mythology.

Bacterial Flagellum: A long hair structure that facilitates the locomotion of bacteria.

Big Bang Theory: The idea that the universe was created by a dense cluster of matter that inflated and expanded into the galaxies and stars that we know today.

Branched-Chain Alpha Keto

Acid Dehydrogenase: An enzyme that catalyzes the catabolism of branched-chain amino acids.

Catabolic Pathways: The breakdown of molecules into simpler ones that will be used in another pathway to make macromolecules.

Caspase Cascade: A group of proteases that help trigger apoptosis. Typically used in cell disassembly and cell death.

Creationist: An individual that believes the universe and living organisms were created by a divine power.

Charles Darwin: English naturalist known for creating the theory of evolution.

Chlorophyll: Green pigments found in plant and algal species.

Collective Origin: The idea that evolution, divine creation, intelligent design, Big Bang theory and all other theories can fit into one bubble as an explanation for the origin of life collectively.

Cytochrome: Proteins that are involved in the electron transport chain. It is composed of a heme and a Fe atom at the core.

Dark Energy: A mysterious force that is responsible for the expansion and acceleration of our universe over time.

Dark Matter: Matter that accounts for about 85% of the universe's matter and holds the universe together. This matter has not been observed and is only an assumption that it exists.

Electron Transport Chain: A series of proteins and molecules embedded along the membrane of a cell. The proteins involved are responsible for the transport of electrons across the membrane.

Evolutionist: An individual in full support of the theories of evolution.

Fore: A small group of people that struggled with a disease called kuru. They live in the Okapa District of the Eastern Highlands Province, Papua New Guinea.

Hubble Space Telescope: A powerful solar powered telescope that can take sharp images and see objects like planets, stars and galaxies.

Inevitable Human Curiosity: The idea that humans seek the origin of life by pre-programmed curiosity and spirituality.

Infinite Derivative Hypothesis: An idea that states life and the universe is infinite. One thing derives from one thing, then the next and the next. The universe and its creator and before the creator are infinite.

Irreducible Complexity: An argument that certain biological systems cannot be reduced to a simpler system. If this is true, then the functionality of that system would cease function (i.e., bacterial flagellum).

Intelligent Design: A belief that the complexities of life cannot be fully supported by scientific theory. Instead, there must have been some type of designer for life.

Kuru: A rare disease caused by a misfolded protein called a prion. Usually found in human brain tissue that has been contaminated.

Law of Conservation: A law that states energy and matter cannot be created nor destroyed. Energy is either conserved or transferred in chemical reactions.

Lac Operon: A genetic regulatory system required for the breakdown of lactose in E. coli.

Microsomal-Ethanol Oxidizing System: A secondary system aside from alcohol dehydrogenase that promotes the conversion of alcohol to acetaldehyde when there is an excessive amount of ethanol consumed.

New Guinea: An island south of the equator and north of the Australian mainland.

Phagocyte: A cell that engulfs other cells and other small particles.

Phosphoglycerate: An important enzyme used in glycolysis that helps produce ATP.

Plastocyanin: Serves a transporter for electrons in the electron transport chain in photosynthesis.

Plastoquinone: A molecule involved in the transport of electrons in the light-dependent reactions of photosynthesis.

Prion: A faulty protein that can change normal proteins into misfolded proteins (abnormal) in the brain. This causes harmful diseases in humans and animals.

Programmed Cell Death: A regulated process for cell death or cell suicide.

Quantum Fluctuations: Random change in the energy at a given point in space.

Ribulose Biphosphate: A substance involved in photosynthesis. It is responsible for the catalyzation of molecules from carbon dioxide.

Stroma: The colorless fluid found in the chloroplast. This is like the matrix in the mitochondria.

The Great Reset: The idea that if the universe were to reset, then a similar outcome might be possible each time.

Theist: An individual that believes in the existence of a deity (i.e., God or gods).

Theology: The study of religious belief and God.

Thylakoid Membrane: This is where photosynthesis occurs. This structure is found in the chloroplasts.

T-Lymphocytes: Cells that derive from stem cells in bone marrow. They help create a protection for the body from infections.

Ubiquitin: A small protein that is used to create proteins and destroy faulty proteins by activating proteases.

Index

A

acetaldehyde · 44
acetaldehyde dehydrogenase · 45
acetate · 45
Alanine · 31
alcohol dehydrogenase · 43
Alcohol dehydrogenase · 46
aldehyde · 44
allolactose · 49
altruism · 72
Amino acids · 31
Andromeda Galaxy · 15
apoptosis · 57
aspartate · 59
ATP · 54

B

bacterial flagellum · 28
branched chain alpha keto dehydrogenase · 30

C

C-AMP-CAP complex · 49
CAP site · 48
caspase · 58
Charles Darwin's evolution theory · 23
chloramine · 36
chlorophyll · 54
Collective Origin · 105
creationist · 89
Cyclic Adenosine Monophosphate · 49
cyclic AMP · 49
cytochrome · 44
Cytochrome C · 59
cytoplasm · 44
Cytosine · 31

D

dark energy · 86
dark matter · 86
dark matter collision theory · 86
dehydration reaction · 37
DNA polymerase · 48
Dr. Michael J. Behe · 28

E

ectoderm · 41
electron transport chain · 54
endoplasmic reticulum · 44
endotherm · 41
Escherichia coli · 47
ethanol · 44
Evolution Theory · 28
evolutionist · 89

F

FAAD · 58
FAS ligand · 58
ferredoxin · 55
Fore · 75

G

galactose · 48
gene regulation · 43
gene transcription · 43
germ layers · 41
glucose · 53
glutamine · 59

golgi apparatus · 27
Great Reset · 74
Guanine · 31

H

Hinduism · 79
Hubble Space Telescope · 15

I

inducer · 49
Inevitable Human Curiosity · 71

K

ketone · 44
kuru · 76

L

Law of Conservation · 97
light-dependent · 54
light-independent · 56

M

macromolecules · 38
maitotoxin · 34
mesoderm · 41
metabolism · 44
microevolutions · 29
Microsomal-Ethanol Oxidizing System · 44
mitochondria · 27
Mompoint Paradox · 20

N

NADH · 44
NADPH · 54
neuro-evolution · 72
nucleus · 29

O

operator · 48
organelles · 29

P

paradox · 18
peptide bonds · 36
peptidyl transferase · 37
phagocyte · 58
Phosphate · 55
Phosphoglycerate · 56
photosynthesis · 53
plastocyanin · 55
plastoquinone · 55
pluripotent cells · 41
prion · 77
promoter · 48
-propyl cyanide · 35
protein alpha-keratin · 36

R

repressor protein · 48
ribosome · 27
Ribulose bisphosphate · 56
rough ER · 27

S

Sodium hypochlorite · 36
spirituality · 82
stomata · 54

stroma · 54

T

T- lymphocytes · 58
thylakoid · 54
Thymine · 31
trans-acetylase · 48

U

Ubiquitin · 61
Uracil · 31

V

valine · 59
Vedas · 79

Z

zinc enzyme · 44

Β

β-Galactosidase · 48

Where Did We Come From?

www.ingramcontent.com/pod-product-compliance
Lightning Source LLC
Chambersburg PA
CBHW070553170426
43201CB00012B/1831